ALLE ZEIT WACH
SJ
1842

Springer Series in Synergetics   Editor: Hermann Haken

Synergetics, an interdisciplinary field of research, is concerned with the cooperation of individual parts of a system that produces macroscopic spatial, temporal or functional structures. It deals with deterministic as well as stochastic processes.

1 **Synergetics** An Introduction 3rd Edition
By H. Haken

2 **Synergetics** A Workshop
Editor: H. Haken

3 **Synergetics** Far from Equilibrium
Editors: A. Pacault, C. Vidal

4 **Structural Stability in Physics**
Editors: W. Güttinger, H. Eikemeier

5 **Pattern Formation by Dynamic Systems and Pattern Recognition**
Editor: H. Haken

6 **Dynamics of Synergetic Systems**
Editor: H. Haken

7 **Problems of Biological Physics**
By L. A. Blumenfeld

8 **Stochastic Nonlinear Systems**
in Physics, Chemistry, and Biology
Editors: L. Arnold, R. Lefever

9 **Numerical Methods in the Study of Critical Phenomena**
Editors: J. Della Dora, J. Demongeot, B. Lacolle

10 **The Kinetic Theory of Electromagnetic Processes** By Yu. L. Klimontovich

11 **Chaos and Order in Nature**
Editor: H. Haken

12 **Nonlinear Phenomena in Chemical Dynamics**
Editors: C. Vidal, A. Pacault

13 **Handbook of Stochastic Methods**
for Physics, Chemistry, and the Natural Sciences
2nd Edition By C. W. Gardiner

14 **Concepts and Models of a Quantitative Sociology** The Dynamics of Interacting Populations By W. Weidlich, G. Haag

15 **Noise-Induced Transitions** Theory and Applications in Physics, Chemistry, and Biology
By W. Horsthemke, R. Lefever

16 **Physics of Bioenergetic Processes**
By L. A. Blumenfeld

17 **Evolution of Order and Chaos** in Physics, Chemistry, and Biology Editor H. Haken

18 **The Fokker-Planck Equation**
By H. Risken

19 **Chemical Oscillations, Waves, and Turbulence**
By Y. Kuramoto

20 **Advanced Synergetics**
By H. Haken

21 **Stochastic Phenomena and Chaotic Behaviour in Complex Systems**
Editor: P. Schuster

22 **Synergetics – From Microscopic to Macroscopic Order** Editor: E. Frehland

23 **Synergetics of the Brain**
Editors: E. Başar, H. Flohr, H. Haken, A. J. Mandell

24 **Chaos and Statistical Methods**
Editor: Y. Kuramoto

25 **Dynamics of Hierarchical Systems**
An Evolutionary Approach By J. S. Nicolis

26 **Self-Organization and Management of Social Systems**
Editors: H. Ulrich, G. J. B. Probst

27 **Non-Equilibrium Dynamics in Chemical Systems** Editors: C. Vidal, A. Pacault

28 **Self-Organization.** Autowaves and Structures Far from Equilibrium
Editor: V. I. Krinsky

29 **Temporal Order**
Editors: L. Rensing, N. I. Jaeger

30 **Dynamical Problems in Soliton Systems**
Editor: S. Takeno

31 **Complex Systems – Operational Approaches** in Neurobiology, Physics, and Computers
Editor: H. Haken

32 **Dimensions and Entropies in Chaotic Systems**
Quantification of Complex Behavior
Editor: G. Mayer-Kress

33 **Selforganization by Nonlinear Irreversible Processes**
Editors: W. Ebeling, H. Ulbricht

Selforganization by Nonlinear Irreversible Processes

Proceedings of the Third International Conference
Kühlungsborn, GDR, March 18–22, 1985

Editors: W. Ebeling and H. Ulbricht

With 78 Figures

Springer-Verlag Berlin Heidelberg New York
London Paris Tokyo

Professor Dr. Werner Ebeling
Sektion Physik, Humboldt-Universität Berlin, Invalidenstraße 42,
DDR-1040 Berlin, GDR

Professor Dr. Heinz Ulbricht
Sektion Physik, Wilhelm-Pieck-Universität, Universitätsplatz 3,
DDR-2500 Rostock, GDR

Series Editor:
Professor Dr. Dr. h. c. Hermann Haken
Institut für Theoretische Physik der Universität Stuttgart, Pfaffenwaldring 57/IV,
D-7000 Stuttgart 80, Fed. Rep. of Germany

ISBN-13: 978-3-642-71006-3 e-ISBN-13: 978-3-642-71004-9
DOI: 10.1007/978-3-642-71004-9

Library of Congress Cataloging-in-Publication Data. Selforganization by nonlinear irreversible processes. (Springer series in synergetics ; 33) Includes index. 1. Self-organizing systems–Congresses. 2. Irreversible process–Congresses. 3. Nonequilibrium thermodynamics–Congresses. 4. Chaotic behavior in systems–Congresses. 5. Stochastic processes–Congresses. I. Ebeling, Werner, 1936-. II. Ulbricht, H. (Heinz) III. International Conference on Irreversible Processes and Dissipative Structures (3rd : 1985 : Ostseebad Kühlungsborn, Germany) IV. Series: Springer series in synergetics ; v. 33. Q325.S47 1986 001.53'3 86-20384

Softcover reprint of the hardcover 1st edition 1986

Offset printing: Druckhaus Beltz, Hemsbach, Bookbinding: J. Schäffer OHG, Grünstadt
2153/3150-543210

Preface

These Proceedings contain invited lectures presented at the third International Conference on "Irreversible Processes and Dissipative Structures" in Kühlungsborn (German Democratic Republic) in March, 1985. These conferences, the first of which was held in Rostock in 1977 and the second in Berlin in 1982, are devoted to the study of irreversible processes far from thermal equilibrium and to the phenomena of selforganization. The meeting in Kühlungsborn brought together some 160 mathematicians, physicists, chemists and biologists from 10 countries, who are all interested in the interdisciplinary field of synergetics.

The main topics of the conference were basic concepts of selforganization and evolution, such as entropy, instabilities, nucleation, dissipative structures, chaos and turbulence. The contributions cover methods from thermodynamics, the theory of dynamic systems, stochastic and statistic theory, the method of Green's functions, the fluctuation-dissipation theorem, etc. Several problems are studied in more detail, e.g., the kinetics of nucleation especially in finite systems, the dynamics of interfaces, reaction-diffusion systems, chemical and biochemical pattern formation and information processing. Further, several contributions are devoted to the development of the concepts of chaos and turbulence.

The editors hope that the contributions collected in this volume will provide some new information about the field of selforganization, which is in full development now.

In the name of the participants, we express sincere thanks to the sponsoring organizations and especially to the Physical Society of the German Democratic Republic and to the Wilhelm-Pieck-University at Rostock for their support of the conference. Our special thanks go to R. Mahnke, H. Engel-Herbert, F. Schweitzer and A. Budde for their help in the preparation of the manuscripts. Last but not least, we thank H. Bahlo, who typed the whole manuscript in a perfect way, and express our sincere thanks to Professor H. Haken for inviting us to include these Proceedings in the *Springer Series in Synergetics.*

Berlin, Rostock, GDR
July 1986

W. Ebeling H. Ulbricht

Preface

Contents

Part I

Basic Concepts

Thermodynamic Aspects of Selforganization

W. Ebeling, H. Engel-Herbert, and H. Herzel

Humboldt-Universität Berlin, Sektion Physik, Invalidenstraße 42, DDR-1040 Berlin, GDR

1. Introduction

Thermodynamics is a rather old discipline of physics, however, it is not oldfashioned. On the contrary, such modern topics as the hot big bang model, the theory of black holes, as well as the theory of biological systems /1,2/, show that thermodynamics goes through a renaissance. Thermodynamics is also intimately related to information theory, a key discipline for the study of selforganization and evolution /3,4/. The very origin of this discipline is closely connected with thermodynamical reasoning, as shown in the fundamental papers of SZILARD (1929), SHANNON (1948) and BRILLOUIN (1956). Thus STRATONOVICH, one of the pioneers of several branches of modern information theory, writes that thermodynamics and statistical physics are the cement which hold together the disciplines forming modern information theory /5/. Besides the informational aspects also the direct consideration of thermodynamic functions and of the entropy production is of much interest for the study of selforganization processes /6,7/. The structures created in the process of selforganization are often called "dissipative structures" /6-8/. Besides this term, which underlines the aspect of dissipation, we shall also use the term "autostructures", which underlines the aspect autonomy /9/. The term autostructure is a generalization of well-known terms as "autooscillations" and "autowaves" /10/.

2. The Concept of Entropy

One of the central concepts of thermodynamics is entropy. There is a close correspondence between entropy and order /11-14/. Let us discuss now several entropy concepts.

(i) Phenomenological thermodynamic entropy (Clausius entropy)
As well known, a system in thermodynamic equilibrium is characterized by a unique state function $S_{eq}(E,X)$ depending on energy and the other extensive parameters of the system. The thermodynamic entropy may also be defined for a certain class of nonequilibrium states /6/. For example, the notion of nonequilibrium thermodynamic entropy has a clear meaning in the field of hydrodynamics, however, the equilibrium parameters change throughout space-time according to the solutions of the hydrodynamic equations. According to BOLTZMANN the equilibrium entropy is maximal

$$S_{eq}(E,X) = \sup S(E,X) . \qquad (2.1)$$

Therefore the lowering of entropy with respect to the equilibrium value taken at the same energy and the other extensive parameters may be taken as a measure of order /12,14/. Let us describe now measuring procedures which would allow us to determine experimen-

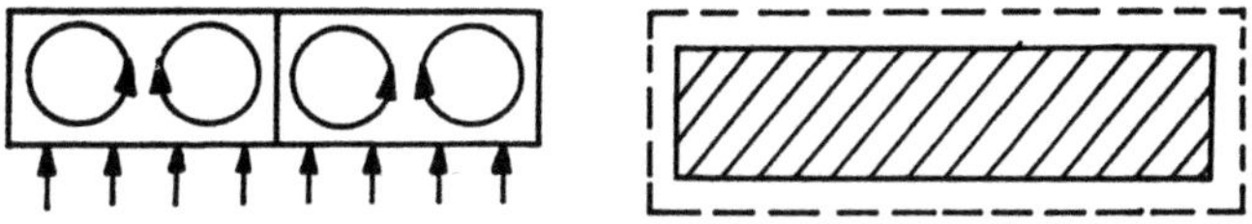

Fig. 1 After total isolation of the system from the environment the dissipative structure (left) disappears and the system relaxes to an equilibrium state (right) characterized by the same values of the energy and the other external parameters

tally the lowering of entropy. In order to compare the order of a given structure with that of the corresponding equilibrium structure we consider the following Gedanken Experiment (see Fig. 1). At a certain moment the system is totally isolated from its surroundings, especially from its pumping sources. Then waiting a sufficiently long time, the system reaches an equilibrium state with the same energy and fixed other extensive parameters. By Boltzmann's principle,

$$\delta S = S_{eq}(E,X) - S(E,X) \qquad (2.2)$$

is always larger zero if the initial state is different from the equilibrium. This, we believe, is a clear operational definition of the thermodynamic measure of order. Let us note that the change is nearly always very small in comparis on to the equilibrium value

$$\zeta = \frac{\delta S}{S_{eq}} \ll 1 . \qquad (2.3)$$

Let us mention further that δS is a smooth function even at bifurcation points.

(ii) Shannon entropy. For a given probability distribution P(x) of states x the Shannon entropy is defined by

$$H = - \int dx\, P(x) \ln P(x) \quad . \qquad (2.4)$$

Here $x = \{x_1,\ldots,x_d\}$ is to be understood as a complete set of order parameters describing the state of the system. For the special case that the state can be identified with the microscopic mechanical state

$$x = \{\vec{q}_1,\vec{q}_2,\ldots,\vec{q}_N,\vec{p}_1,\vec{p}_2,\ldots\vec{p}_N\} , \qquad (2.5)$$

the Shannon entropy is related to the statistical entropy proposed by BOLTZMANN: $S = k_B H$ (microstate). Therefore the statistical entropy corresponds to the information which is necessary to specify the microstate. Let us underline that for other possible choices of the state space, i.e. for reduced descriptions, the Shannon entropy is not necessarily related to the statistical entropy.

In terms of the distribution function the entropy after BOLTZMANN is given by

$$S = - k_B \int \frac{d\vec{p}d\vec{q}}{(2\pi\hbar)^3} f(\vec{p},\vec{q}) \ln f(\vec{p},\vec{q}) , \qquad (2.6)$$

where the distribution function is normalized as

$$\int \frac{d\vec{p}d\vec{q}}{(2\pi\hbar)^3} f(\vec{p},\vec{q}) = N . \qquad (2.7)$$

Strictly speaking, formula (2.6) is valid only for noninteracting systems. However, in many cases selforganization is not connected with a change of spatial correlations but mainly with a change of velocities. Therefore we will apply (2.6) even to selforganization phenomena in liquids. The generalization of (2.6) to interacting systems is after GIBBS given by

$$S = -k_B \int \frac{d\vec{q}_1 \ldots d\vec{p}_N}{(2\pi\hbar)^{3N}} \varrho(\vec{q}_1,\ldots,\vec{p}_N) \ln \varrho(\vec{q}_1,\ldots,\vec{p}_N) . \tag{2.8}$$

Here $\varrho(\vec{q}_1,\ldots,\vec{p}_N)$ is the "coarse-grained" probability density in the full phase space. The Boltzmann approximation (2.6) corresponds to a factorization of ϱ.

Statistical and thermodynamic entropy certainly coincide for macroscopic systems in thermodynamic equilibrium. BOLTZMANN and GIBBS extended the notion of statistical entropy to nonequilibrium states. For the one particle distribution function f(q,p,t) BOLTZMANN proved the well-known H-theorem to give a statistical interpretation of the second law of thermodynamics.

(iii) Algorithmic entropy. For simplicity let us consider only a special type of structures consisting of strings of letters in a certain alphabet. $A_1 \ldots A_\lambda$. For example DNA-molecules are strings (words) in the alphabet A,C,G,T /15/. A special string is e.g.

CAGTACAGTA ... ATCA .

KOLMOGOROV and CHAITIN introduced in 1965 the concept of complexity of a sequence expressing the minimal length of programs which are able to generate this sequence.

(iv) Mean entropy per symbol for sequences of letters. Considering strings of symbols, this entropy is defined as

$$H = \lim_{r \to \infty} (H_{r+1} - H_r) , \tag{2.9}$$

where the H_r denote the Shannon entropies for r-words /7/. The minimal value of H is zero, corresponding to regularity and complete predictability of the following letter in a sequence.

(v) Kolmogorov entropy for dynamical systems. Using partitions $\{A\}$ of the phase space and symbolic representations of the trajectory by strings of letters, a Kolmogorov entropy may be defined by /16-18/

$$K = \sup_{\{A\}} H(\{A\}) . \tag{2.10}$$

The minimal value of K is zero, which means regularity or periodicity of the motion which allows predictability of the future. Positive values of K mean chaotic motion and restricted predictability /19/.

The list of entropy definitions given above is not complete. Other measures of order are connected with breaking of symmetry or with the increase of correlations /13/.

3. Thermodynamic Balances and the Driving Forces of Selforganization

Let us consider open nonequilibrium systems exchanging energy E, entropy S and matter N_k with the surroundings. The balance equations for these quantities read

$$\frac{dE}{dt} = \frac{d_iE}{dt} + \frac{d_eE}{dt} , \tag{3.1}$$

$$\frac{dS}{dt} = \frac{d_iS}{dt} + \frac{d_eS}{dt} \quad , \tag{3.2}$$

$$\frac{dN_k}{dt} = \frac{d_iN_k}{dt} + \frac{d_eN_k}{dt} \quad . \tag{3.3}$$

The production parts (internal contributions) are given by the fundamental laws of physics and chemistry.

First law: $$\frac{d_iE}{dt} = 0 \quad . \tag{3.4}$$

Second law: $$P = \frac{d_iS}{dt} > 0 \quad . \tag{3.5}$$

Chemical laws: $$\frac{d_iN_k}{dt} = f_k(N_1, \ldots, N_k) \quad . \tag{3.6}$$

Here the functions f_k express the reaction velocity as a function of the mole numbers (particle numbers) $N_1, N_2, \ldots, N_n$. Selforganization requires that the export of entropy exceeds the import (Fig. 2). Often there are conditions where the processes are isothermal. Then the free energy

$$F = E - TS \tag{3.7}$$

is the basic quantity with the balance

$$\frac{dF}{dt} = - T\frac{d_iS}{dt} + \frac{d_eE}{dt} - T\frac{d_eS}{dt} \quad . \tag{3.8}$$

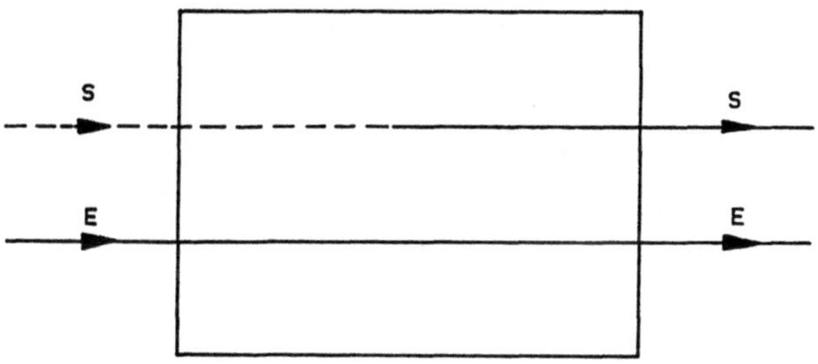

Fig. 2 Scheme of the energy and entropy flows in selforganizing systems

Therefore the flow of free energy is with (3.8) and (3.4)

$$\Phi_F = \frac{dE}{dt} - T\frac{d_eS}{dt} \quad . \tag{3.9}$$

Let us split now the free energy flow in a part which is pumped into the system Φ_p, a part which is given back unchanged Φ_a and a part which is transformed into another possibly more useful form Φ_n which is utilized somehow /20/

$$\Phi_F = \Phi_p - \Phi_a - \Phi_n \quad . \tag{3.10}$$

Then the total free energy balance resulting from (3.9-3.10) reads

$$\frac{dF}{dt} = \Phi_p - \Phi_a - \Phi_n + T\,P \;,\; P = \frac{d_iS}{dt} \;. \tag{3.11}$$

For stationary states follows from (3.2), (3.5) and (3.11)

$$\Phi_p = \Phi_a + \Phi_n + T\,P \;. \tag{3.12}$$

The efficiency of the dissipative structures considered as an energy-transforming device may be defined as /20/

$$\eta = \frac{\Phi_n}{\Phi_p - \Phi_a} = \frac{\Phi_n}{\Phi_n + TP} < 1. \tag{3.13}$$

The inequality follows from the second law (3.5).

Up to now we have considered the dissipative structure as a kind of black box with certain inflow and outflow of energy. Now let us consider from the thermodynamical point of view what happens in the interior of the system. Evidently the dissipative structure is a kind of energy-transforming engine which uses a part of the energy inflow to produce a new form of energy which is of higher thermodynamic value (i.e. lower entropy); examples are the mechanical energy in Benard cells, the electrical energy in nerve membranes and the concentration-gradients (activity gradients) in chemical dissipative structures. We notice in this connection that concentration difference energy (CDE) is a high-valued energy (due to the low corresponding entropy). Therefore a chemical dissipative structure with big concentration differences represents also a depot of high-valued energy. If a part of this energy will be utilized for other purpose the system will fill up again its depot of valuable free energy by transforming some part of the energy inflow.

In recent papers /21/ the efficiency of a nonequilibrium battery (Fig. 3) based on symmetry-breaking structures in two-box systems with autocatalytic ionic reactions of the type

$$A \rightleftharpoons X_1^+ + X_2^- \;;\; B + 2X_1^+ \rightleftharpoons C + 3X_1^+ + X_2^- \tag{3.14}$$

has been calculated. Unsymmetrical ion distributions lead to a voltage which may drive a current in an external circuit. Using realistic parameter values, we have obtained the following result for the voltage, the power, the losses and the efficiency:

$$\varphi = 39.8 \text{ V} \;, \qquad \Phi_n = 7.3 \cdot 10^{-7} \text{ W}$$

$$TP = 6 \cdot 10^{-3} \text{ W}, \qquad \eta = 1.1 \cdot 10^{-4} \;.$$

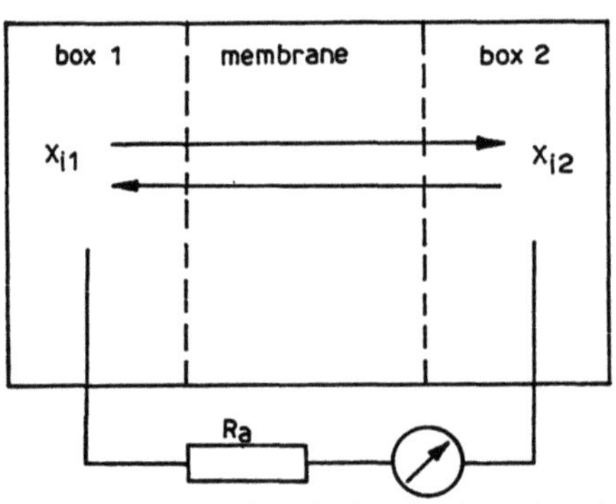

Fig. 3 An electrical dissipative structure: the nonequilibrium battery

As we see, the efficiency of dissipative structures (autostructures) is quite low. The low efficiency of the dissipative structures discussed above (with respect to energy transformations) is due to losses which are unavoidable under far from equilibrium conditions. Let us note, however, that the "autostructure engines", have the advantage of great conceptual simplicity, i.e. they are formed spontaneously, they do not need inventors and engineers. This may be the main reason for the broad applications of autostructure engines in living systems; they have been tested in the process of natural evolution /7/ and now they are still working despite a relatively low efficiency. Possible applications of autostructure engines for special technical purposes should be a subject of further research.

Let us consider now the driving forces of selforganization in nature and in technical systems.

We have shown above that selforganization is necessarily connected with the possibility of exporting entropy to the external world. In other words, selforganizing systems need an input of high-valued energy and at the same time an output of low-valued energy. In the interior of selforganizing systems a depot of high-valued energy of another form is observed. The evolution processes on our planet are mainly pumped by the "photon mill" with the three levels sun/earth/background radiation. On the cosmic scale, the general strategy of evolution is the formation of islands of order on a sea of disorder represented by the background radiation. Let us consider now this mechanism of selforganization on the cosmic scale /7,9/. The standard model for the evolution of the metagalaxis tells us that about 10^5 years after the "big bang" the metagalaxis was decomposed into two nearly independent systems (Fig. 4):

(i) The matter with rest mass
(ii) The radiation field.

This happened at a temperature of about $10^3 - 10^4$ K when the electrons were captured by the nuclei and the heavy matter formed a neutral gas. Due to the adiabatic expansion the radiation field was all the time decreasing in temperature. Nowadays the universe is homogeneously filled with photons in thermal equilibrium at a temperature of 3 K. We shall see that the sea of 3 K photons is the waste of our metagalaxis which takes over the entropy from the selforganizing systems. Where do the evolutionary possibilities of our metagalaxy come from? We know that the neutral gas lost the homogeneity due to gravitational instabilities and formed stars with temperatures of about 10^4 K at the surface and planets with about 10^2 K at the surface. The "falling down" of the photons from the surface of the stars to the surface of the planets and from there to the sea of background radiation is the "mill" which drives the "wheel of evolution". The total energy flow which reaches the earth from the sun is

$$\frac{d_eE}{dt} = 1.2 \cdot 10^{17}\ \mathrm{W} \tag{3.15}$$

and the same quantity is exported again, however with a much lower radiation temperature of about 260 K (see Fig. 4). Therefore we get for the total entropy export /7/

$$\frac{d_eS}{dt} = \frac{4}{3} \cdot 1.2 \cdot 10^{17}\,\mathrm{W} \left(\frac{1}{5800\,\mathrm{K}} - \frac{1}{260\,\mathrm{K}}\right) = -6 \cdot 10^{14}\,\frac{\mathrm{W}}{\mathrm{K}} \; . \tag{3.16}$$

This gives on average per m^2 the entropy flow density

$$J_s = -1\,\frac{\mathrm{W}}{\mathrm{m}^2\mathrm{K}} \; . \tag{3.17}$$

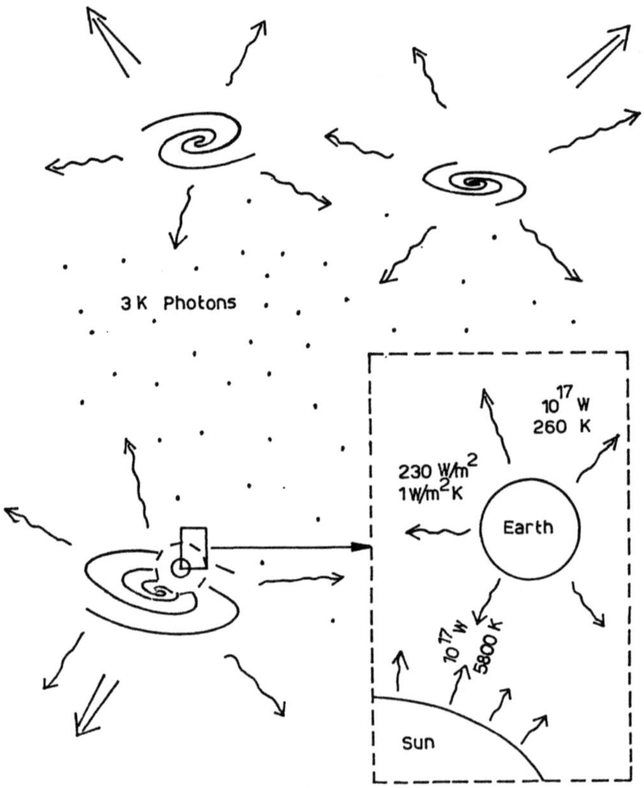

Fig. 4 Scheme of the energy flow in an expanding universe. In the window: The energy and entropy balance of the earth

This number, which is of fundamental importance, expresses the ability of our planet with respect to selforganization and evolution giving an upper limit to the average entropy export. We may be sure that this number is deeply connected not only with the possibility but also with the speed and the results of the evolution of life on earth.

As we have shown above, the driving force of selforganization on the cosmic scale is due to the strong nonequilibrium at the beginning of the evolution of our present universe. Furtheron, an essential role play the gravitational instabilities of uniform matter distributions, which are responsible for the island structure of our universe and for the photon mill mechanism.

4. Oscillator Systems

Sustained oscillations are typical nonequilibrium structures which appear due to a spontaneous breaking of time symmetry if the feedback strength exceeds some critical level. In order to demonstrate that the generation of sustained oscillations is connected with an entropy lowering, as explained above, we consider an ideal gas of N nonlinear oscillators with the equations of motion

$$\ddot{q} + \varepsilon\left(\frac{1}{2}u_1 + \frac{2}{3}u_3\dot{q}^2 + \frac{3}{5}u_5\dot{q}^4\right)\dot{q} + \omega_0^2 q = \sqrt{\varepsilon D}\ \zeta(t) ; \qquad (4.1)$$

q denotes the coordinate of the oscillator and u_1, u_3, u_5 are parameters of the nonlinear damping characteristics. D is the intensity of the fluctuations caused by the collisions between the oscillators. The fluctuations will be approximated by δ-correlated white noise

$$\langle \xi(t) \rangle = 0 \ , \quad \langle \xi(t+\tau)\xi(t) \rangle = \delta(\tau) \ . \tag{4.2}$$

It is convenient to decompose u_1 into two parts

$$u_1 = \gamma - \delta , \tag{4.3}$$

where γ is the linear damping coefficient which is due to molecular interactions and δ represents the feedback strength. If δ is increased beyond the threshold value

$$\delta_{cr} = \begin{cases} \gamma & \text{for soft mechanism of excitation (if } u_3 \geq 0) \\ \gamma - \dfrac{1}{3}\dfrac{u_3^2}{u_5} & \text{for hard mechanism of excitation (if } u_3 < 0) \end{cases} \tag{4.4}$$

sustained oscillations are generated. We are interested in the behaviour of the entropy during the transition.

Using time-scale arguments valid for sufficiently small ε we obtain the following stationary distribution function of the oscillator energy $H = (\omega_0^2 q^2 + q^2)/2$ /23/

$$f^0(H) = \frac{1}{Z} \exp\left[-\frac{1}{D}(u_1 H + u_3 H^2 + u_5 H^3)\right] . \tag{4.5}$$

According to the general relation (2.6), from (4.5) follows

$$\frac{1}{3Nk_B} S = \ln Z + \frac{u_1}{D}\langle H \rangle + \frac{u_3}{D}\langle H^2 \rangle + \frac{u_5}{D}\langle H^3 \rangle + \text{const} , \tag{4.6}$$

where

$$Z = \int_0^\infty \exp\left[-\frac{1}{D}(u_1 H + u_2 H^2 + u_3 H^3)\right] ,$$

$$\langle H^n \rangle = \int_0^\infty H^n f^0(H) dH \ , \quad n = 1,2,\ldots \tag{4.7}$$

denote the normalization constant and the moments of the stationary energy distribution respectively.

First we consider the soft excitation of oscillations. Without loss of generality we may put

$$u_5 = 0, \quad u_3 > 0 \ . \tag{4.8}$$

For any value of δ we have to determine the intensity of the fluctuations $D(\delta)$ in accordance with the condition that the average energy remains constant during the transition

$$\langle H \rangle = \text{const.} \tag{4.9}$$

Furthermore we choose

$$D(\delta = 0) = \gamma k_B T \tag{4.10}$$

to ensure that the system relaxes to thermodynamic equilibrium if the active nonlinearities vanish ($\delta = u_3 = 0$). Then we find from (4.6-4.8), using

$$\int_0^\infty H^n e^{-\alpha H^2-\beta H}\, dH = \frac{(-1)^n}{2}\sqrt{\frac{\pi}{\alpha}}\frac{\partial^n}{\partial\beta^n} F\left(\frac{\beta}{2\alpha}\right),$$
$$F(y) = \frac{2}{\sqrt{\pi}} e^{y^2}\int_y^\infty e^{-t^2}dt , \tag{4.11}$$

the following expression for the entropy

$$\frac{1}{3Nk_B} S = \ln\left[\frac{1}{2}\frac{\pi}{\beta} F\left(\frac{\beta}{2\sqrt{\alpha}}\right)\right] + \frac{\beta}{2\pi\alpha F(\beta/2\sqrt{\alpha})} - \frac{\beta^2}{4\alpha} + \frac{1}{2} + const,$$
$$\beta \equiv \frac{\gamma-\delta}{D}, \quad \alpha \equiv \frac{u_3}{D} . \tag{4.12}$$

The function $D(\delta)$ has to be determined for given value of $\langle H\rangle$ from (4.9), i.e. from

$$\langle H\rangle = \frac{\sqrt{\frac{D}{\pi u_3}}}{F\left(\frac{\gamma-\delta}{2\sqrt{Du_3}}\right)} - \frac{\gamma-\delta}{2u_3} . \tag{4.13}$$

KLIMONTOVICH has shown /12/ that for fixed $\langle H\rangle$ the entropy at the threshold exceeds the entropy in the regime of fully developed oscillations and is smaller than in the regime without feedback. Numerical calculations /23/ suggested that the entropy is a monotonously decreasing function of the feedback strength. This can be shown analytically. Using (4.5-4.8) one verifies the relation

$$D^4\left(\frac{\partial S}{\partial\delta}\right)_{\langle H\rangle}\left(\frac{\partial S}{\partial\delta}\right)_D = u_3^2\ [\langle(H-\langle H\rangle)(H^2-\langle H^2\rangle)\rangle^2 - \langle(H-\langle H\rangle)^2\rangle\ \langle(H^2-\langle H^2\rangle)^2\rangle] , \tag{4.14}$$

which connects the derivatives of the entropy S with respect to the feedback strength δ at constant average oscillator energy and at constant noise intensity D respectively. Since

$$D\left(\frac{\partial S}{\partial\delta}\right)_D = \beta\ \langle H(H-\langle H\rangle)\rangle + \alpha\langle H^2(H-\langle H\rangle)\rangle$$
$$= \frac{\left(\frac{\beta^2}{4\alpha}+\frac{1}{2}\right)F\left(\frac{\beta}{2\sqrt{\alpha}}\right) - \frac{\beta}{2\sqrt{\pi\alpha}}}{\sqrt{\pi\alpha}\, F^2\left(\frac{\beta}{2\sqrt{\alpha}}\right)} = \frac{\langle H^2\rangle}{Z} \tag{4.15}$$

is strictly positive we conclude

$$\left(\frac{\partial S}{\partial\delta}\right)_{\langle H\rangle} \leq 0 . \tag{4.16}$$

The hard transition to sustained oscillations ($u_5 > 0$) is of interest because it is governed by two independent parameters u_1 and u_3. In the u_1, u_3-plane we distinguish three regions of qualitatively different behaviour (Fig. 5):

(i) The region of damped motion $u_1 > 0$, $u_3 \geq -\sqrt{3u_1u_5}$,

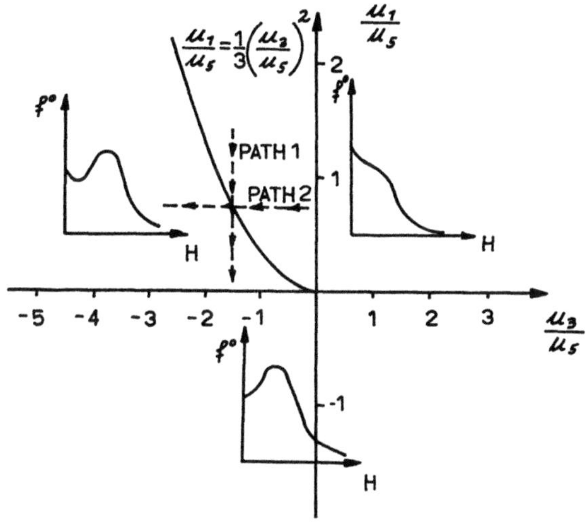

Fig. 5 The bifurcation diagram for the nonlinear oscillator (4.1 - 4.3) with the shape of the energy distribution in the three regions of qualitatively different behaviour

(ii) the oscillatory region with one stable limit cycle: $u_1 < 0$,

(iii) the oscillatory region where a stable and an unstable limit cycle coexist: $u_1 \geqslant 0$, $u_3 < -\sqrt{3u_1u_5}$.

Here we only briefly discuss some results obtained numerically for the hard transition according to path 1 and path 2 in Fig. 5 /24/. Along path 1 ($u_3 = -30$, $u_5 = 20$) the noise intensity was determined for given values of δ in correspondence with the condition (4.9) where $\langle H \rangle = 1.0$ was chosen. This effective noise intensity $D(\delta)$ decreases with increasing δ . Calculating S using the corresponding values for D again an entropy lowering during the transition was observed. Quite the same behaviour takes place if the oscillatory region is entered along path 2 ($u_1 = 15$, $u_3 = 2$, $\langle H \rangle = 0.7$). For the derivative of the entropy with respect to the parameter governing the transition yields

$$\left(\frac{\partial S}{\partial \delta}\right)_{\langle H\rangle, u_3} < 0 \quad \text{for path 1 and} \quad \left(\frac{\partial S}{\partial u_3}\right)_{\langle H\rangle, \delta} > 0 \quad \text{for path 2.} \tag{4.17}$$

Summarizing, in the situations considered here the generation of sustained oscillations is connected with entropy lowering so far as the mean energy of the oscillations does not change during the transition.

5. Turbulence Flows

We describe the flow by the one-particle distribution function $\tilde{f}(\vec{r},\vec{p},t)$ and take into account that $\tilde{f}$ will be a random function /11/

$$\tilde{f} = f^0 + \tilde{f}^1 , \tag{5.1}$$

where

$$f^0 = \frac{nh^3}{(2\pi m k_B T)^{3/2}} \quad \exp\left[-\frac{(\vec{p}-m\vec{u})^2}{2 m k_B T}\right] \tag{5.2}$$

represents the local Maxwell distribution function ($\vec{v}=\vec{p}/m$ - velocity of the particles, $\tilde{u}$ - hydrodynamic velocity) and the nonequilibrium part $\tilde{f}^1 = f^1 + \delta\tilde{f}$ includes the fluctuations of the distribution function $\delta\tilde{f}$ ($\langle\delta\tilde{f}\rangle = 0$). We assume that $\tilde{f}$ only weakly deviates from f^0. Under this assumption we get for the entropy according to (2.6)

$$\tilde{S} = - k_B \int \frac{d\vec{r}d\vec{p}}{h^3} (f^0 \ln f^0 + \tilde{f}^1 \ln f^0 + \frac{1}{2} \frac{(\tilde{f}^1)^2}{f^0} + \ldots). \tag{5.3}$$

The first term in (5.3) represents the equilibrium contribution

$$S^0 = Nk_B \left\{\frac{3}{2} - \ln\left[\frac{nh^3}{(2\pi m k_B T)^{3/2}}\right]\right\} ; \tag{5.4}$$

f^1 may be determined by an expansion in Hermitian polynomials. The hydrodynamic approximation yields

$$\tilde{f} = f^0 \left\{1 + \frac{1}{2\varrho}\left(\frac{m}{k_B T}\right)^2 \tilde{\Pi}_{ij}\, \delta\tilde{v}_i \delta\tilde{v}_j + \frac{m(\delta\tilde{v}_i\, \tilde{q}_i)}{p_{id}\, k_B T}\left[\frac{m(\delta\tilde{v}_i\, \delta\tilde{v}_i)}{5 k_B T} - 1\right] + \ldots\right\}, \tag{5.5}$$

where $\delta\tilde{v}_i = v_i - \tilde{u}_i$, $p_{id} = nk_B T$, ϱ is the density and

$$\tilde{\Pi}_{ij} = -\eta\left(\frac{\partial \tilde{u}_i}{\partial r_j} + \frac{\partial \tilde{u}_j}{\partial r_i}\right),$$

$$\tilde{\vec{q}} = -\lambda \operatorname{grad} \tilde{T} \tag{5.6}$$

denote the fluctuating stress tensor and the heat flow respectively. With (5.5) the second term in (5.3) vanishes and for the entropy follows

$$\langle\tilde{S}\rangle - S^0 = -k_B \int d\vec{r} \left(\frac{1}{4p_{id}^2} \langle \tilde{\Pi}_{ij}\, \tilde{\Pi}_{ij}\rangle + \frac{\varrho}{p_{id}^3} \langle \tilde{q}_i\, \tilde{q}_i\rangle\right). \tag{5.7}$$

The velocity field $\tilde{\vec{u}}$ and the temperature distribution $\tilde{T}$ have to be determined from the hydrodynamical equations considered as equations for fluctuating quantities.

The condition of constant energy reads, for our quasi-ideal system without potential energy,

$$\int d\vec{r}\, d\vec{p}\, \frac{\vec{p}^2}{2m} f^0 = \int d\vec{r}\, d\vec{p}\, \frac{\vec{p}^2}{2m} \tilde{f} . \tag{5.8}$$

Using (5.9), we define the temperature of the isoenergetic laminar or turbulent flow

$$T = T^0 - \frac{m}{3k_B NV} \int d\vec{r} \langle(\tilde{\vec{u}})^2\rangle. \tag{5.9}$$

T^0 is the temperature reached after the relaxation to equilibrium if the system was isolated from the surroundings (see Fig. 1). We remark that (5.7) may be written in the equivalent form

$$\langle\tilde{S}\rangle = S^0(T) - \frac{\eta}{2p_{id}} P_\eta - \frac{\varrho\lambda T}{p_{id}^2} P_\lambda , \tag{5.10}$$

where

$$P_\eta = \int d\vec{r}\, \frac{\langle \tilde{\Pi}_{ij} \tilde{\Pi}_{ij} \rangle}{2\eta T} \quad \text{and} \quad P_\lambda = \int d\vec{r}\, \frac{\langle \tilde{q}_i \tilde{q}_i \rangle}{\lambda T^2} \tag{5.11}$$

denote the part of the entropy production caused by viscous forces and by heat conduction respectively. $\eta/2p_{id}$ and $\varrho\lambda T/(p_{id})^2$ may be interpreted as relaxation times which characterize the decay of elastic and thermal deformations in the liquid. In general these two nonequilibrium contributions are small compared with S^0 and will be neglected in the following calculation. In this approximation we find for the entropy lowering

$$\delta S = \langle \tilde{S} \rangle - S^0 \cong \frac{3}{2} N k_B \ln \frac{T}{T^0}$$

$$= \frac{3}{2} N k_B \ln\left(1 - \frac{m}{3k_B N T} \int \frac{d\vec{r}}{V} \langle (\tilde{\vec{u}})^2 \rangle \right) < 0 \, . \tag{5.12}$$

Finally we briefly discuss the flow in a pipe of radius a driven by the pressure gradient $\Delta p/l$. To determine the effective temperature of the flow we use the average turbulent velocity profile /22/

$$u(r) = v^* R^0_{cr} \ln\left[1 + \frac{R^*}{2R^0_{cr}} (1 - \bar{r}^2)\right] \, , \quad \bar{r} \equiv \frac{r}{a} \tag{5.13}$$

(r - radial coordinate, z-axis along the axis of the pipe). The Reynolds number R^* is related to the dynamic velocity v^* via $R^* = v^* a/\nu$ (ν - kinematic viscosity) where v^* is defined by $v^* = (\Delta p a/2\varrho l)^{1/2}$. R^0_{cr} characterizes the dimension of the smallest eddies in the flow. For $R^*/R^0_{cr} \ll 1$ expression (5.13) transforms to the Poiseuille profile.

The turbulent pulsations will be approximated in the following way

$$\tilde{u}_i = u_i + \delta u_i \, , \quad \langle (\delta u_x)^2 \rangle \cong \langle (\delta u_y)^2 \rangle \cong \langle (\delta u_z)^2 \rangle \cong v^{*2} \, . \tag{5.14}$$

The results obtained from (5.12-5.14) are presented in Fig. 6. As shown there, the lowering of entropy increases with the Reynolds number $R = ua/\nu$ where u denotes the average turbulent velocity of the flow. Let us repeat here again that δS corresponds to the increase in entropy if the pipe is at certain moment isolated from the environment and relaxes to thermal equilibrium. In other words δS is the difference of entropy between the isoenergetic equilibrium system (which will be reached after relaxation) and the actual flow in the pipe. Looking at Fig. 6, we see that at given Reynolds number the entropy lowering for a laminar flow, which is unstable at overcritical Reynolds numbers, is higher than for the corresponding turbulent flow. In this sense the laminar flow is more ordered than the corresponding turbulent flow at the same Reynolds number. However, if we compare laminar flows at low Reynolds numbers with turbulent flows at high Reynolds numbers we see that in this respect the turbulent flow is more ordered (measured by the entropy lowering). This is due to the fact that with increasing Reynolds number a larger percentage of energy is concentrated in the collective modes of hydrodynamic motion consisting of the regular part of the flow and the turbulent pulsations.

RUELLE and TAKENS have suggested that hydrodynamic turbulence corresponds to solutions of finite dimensional deterministic dyna-

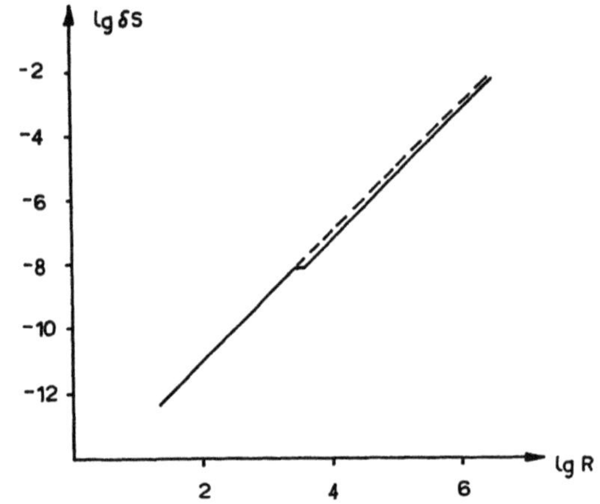

Fig. 6 Entropy lowering δS for the Poiseuille flow against the Reynolds number R. The dashed line corresponds to the unstable laminar regime above $R_{cr} \sim 10^3$

mics asymptotic to a strange attractor /26/. In agreement with this concept for weak turbulence in TAYLOR-COUETTE flow, a positive largest Ljapunov exponent and a positive Kolmogorov entropy were obtained from experimental data /27/. Recently upper bounds on the Kolmogorov entropy and the Hausdorff dimension of the strange attractor of the Navier-Stokes equation were determined /28/. Both quantities increase with the Reynolds number. The growth of the attractor dimension with R indicates an increase of the number of excited degrees of freedom N_t in the flow. For a fully developed turbulent flow the old estimate by KOLMOGOROV was recovered that N_t is of order $R^{9/4}$.

6. Conclusions

We have shown that the thermodynamic aspects and especially the concept of entropy are of relevance for the understanding of selforganization. Several entropy concepts, such as the thermodynamic entropy (Clausius entropy), the statistical entropy (Boltzmann-Gibbs entropy), the Shannon entropy, the algorithmic entropy and the Kolmogorov entropy have been discussed.

This work mainly concentrates on the Clausius entropy and the Boltzmann-Gibbs entropy, which are identical if the proper distribution functions are introduced into the Boltzmann-Gibbs formula. Explicit calculations are given for the entropy lowering for two examples: (i) the gas of excited oscillators and (ii) laminar and turbulent Poiseuille flow.

It was shown that under the condition of constant energy the entropy lowering increases with the feedback strength for the oscillator gas and with the Reynolds number for the flow. For both systems, judging from the entropy, the stationary nonequilibrium state is more ordered than the equilibrium state of the same energy.

Roughly speaking, the entropy is given by the logarithm of the available phase volume. To avoid confusion, let us underline that the available volume in the phase space is to be understood in the sense of a coarse-graining procedure /29/. Among all states of given energy the entropy is maximal in thermodynamic equilibrium. In the equilibrium state the distribution function covers the whole energy shell in the phase space. Stationary nonequilibrium states correspond to attractors on the energy shell. The formation of attractors reduces the available phase volume. This is reflected by the entropy lowering in comparison with the isoenergetic equilibrium state. If the system in a stationary nonequilibrium state is isolated from its surroundings, it remains on the same energy shell, but the attractor disappears and after a certain time necessary to reach the equilibrium state again the whole energy shell will be occupied by the distribution function. During the relaxation to equilibrium the entropy increases with the logarithm of the available phase volume.

There exists another possible terminology for the description of selforganization. We may say that the filling up of the energy shell with constant density corresponds to the lowest value of the given energy (measured by the entropy). The formation of attractors means that the value of the energy increases, since attractors in the phase space correspond to ordered motions of the particles. Selforganization is connection with the appearance of attractors in the phase space, and this requires the pumping of the system with high-valued energy. It is shown that in our universe the main source of high-valued energy is the photon mill, i.e. the falling down of the photons from the temperature level 10^4 K at the surface of the stars to the 10^1 K level of the background radiation homogeneously distributed in the cosmos.

References

1. I. Lamprecht, A.J. Zotin (Eds.): Thermodynamics and Kinetics of Biological Processes (De Gruyter, Berlin, New York 1982)
2. I. Lamprecht, A.J. Zotin (Eds.): Thermodynamics and Regulation (De Gruyter, Berlin New York 1985)
3. H. Haken: Advanced Synergetics (Springer, Berlin Heidelberg New York 1983)
4. H. Haken: Towards a Dynamic Information Theory, in Ref. 1
5. R.L. Stratonovich: Theory of Information (Mir, Moscow 1977 (in Russian))
6. P. Glansdorff, I. Prigogine: Thermodynamics of Structure, Stability and Fluctuations (Wiley, New York London Toronto Sydney 1971), G. Nicolis, I. Prigogine: Selforganization in Nonequilibrium Systems (Wiley, New York London Sydney Toronto 1977)
7. W. Ebeling, R. Feistel: Physik der Selbstorganisation und Evolution (Akademie-Verlag, Berlin 1982)
8. V.A. Vasiliev, Ju.M. Romanovsky, D.S. Chernavsky, V.G. Yakhno: Autowave Processes in Distributed Kinetic Systems (Verlag der Wissenschaften, Berlin in press)
9. W. Ebeling: Biomed.Biochim.Acta 44, 831 (1985)
10. V.J. Krinsky (Ed.): Autowaves and Structures far from Equilibrium Springer Series in Synergetics vol. 28 (Springer, Berlin Heidelberg New York 1984)
11. Ju.L. Klimontovich: Statistical Physics (Nauka, Moscow 1982 (in Russian))
12. Ju.L. Klimontovich: Pisma Zh.Tekh.Fiz. 9, 1412 (1983); 10, 50 (1984)
13. G. Careri: Order and Disorder in Matter (Benjamin/Cummings, Menlo Park 1984)
14. W. Ebeling, Yu.L. Klimontovich: Selforganization and Turbulence in Liquids (Teubner, Leipzig 1984)
15. W. Ebeling, M.A. Jimenez-Montano: Math.Biosci. 52, 53 (1980)
16. J.D. Farmer: Z.Naturf. 37a, 1304 (1982)
17. J.-P. Eckmann, D. Ruelle: Rev.Mod.Phys. 57, 617 (1985)
18. G. Györgi, P. Szepfalusy: Phys.Rev. 31A, 3477 (1985)
19. H. Herzel, W. Ebeling: Phys.Lett. 111A, 1 (1985)
20. R. Feistel, W. Ebeling: Exp.Technik der Physik 32, 1 (1984)

21. W. Ebeling, U. Feudel: Ann.Physik (Leipzig) 40, 68 (1983); U. Feudel, W. Ebeling, R. Feistel: Ann.Physik (Leipzig) 41, 100 (1984)
22. W. Ebeling, H. Engel-Herbert: Physica 104A, 378 (1980); Adv. Mechanics 5, 41 (1982)
23. W. Ebeling, H. Herzel, H. Engel-Herbert: Z.Phys.Chem. (Leipzig) 266, 253 (1985)
24. H. Engel-Herbert, M. Schumann: Ann.Physik (Leipzig) to appear
25. Yu.L. Klimontovich, H. Engel-Herbert: Zh.Tekhn.Physic (Moscow) 54, 440 (1984)
26. D. Ruelle, F. Takens: Commun.Math.Phys. 20, 167 (1971)
27. A. Brandstätter, J. Swift, H.L. Swinney, A. Wolf, J.D. Farmer, E. Jen, P.J. Crutchfield: Phys.Rev.Lett. 51, 1442 (1983)
28. D. Ruelle: Physica Scripta 9, 147 (1985)
29. G.M. Zaslavski: Stochasticity of Dynamical Systems (Nauka, Moscow 1984 (in Russian))

Entropy in Time Averaging

Y.L. Klimontovich

Moscow State University, Department of Physics, SU-117234 Moscow, USSR

1. Time Averaging

Due to the work of Ch. DARWIN, selforganization in nature became a central problem of science. Only recently (1977-1978), however, the basis of a theory of selforganization was formulated by I. PRIGOGINE, H. HAKEN and others. In many cases selforganization can be understood as a succession of non-equilibrium phase transitions taking place if one or several control parameters are varied. In this connection the question arises how to compare the degree of order and disorder or chaos respectively of the different states of the system under consideration.

Certainly the degree of order can be defined using different criteria (cf. the contribution of W. EBELING et al. in this book). One possible criterion is the comparison of the values of the Boltzmann-Gibbs entropy corresponding to different values of the control parameter but referring to the same value of the mean energy. Whenever the so calculated entropy decreases selforganization occurs. We have called this hypothesis "S-Theorem".

If there are several control parameters, different paths of evolution are possible, and one can look for optimal conditions for selforganization.

The calculation of the entropy with the help of the formulae of Boltzmann and Gibbs is based on the averaging procedure in phase space. However, in numerical and real experiments it is often more convenient to use time averages and the corresponding distribution functions. This allows to determine the wanted statistical characteristic using a realization of the stochastic process.

We consider an arbitrary system of stochastic dissipative equations and denote by X the complete set of variables. A(X) is an arbitrary function of X. From the solution X(t) the corresponding realization $A(X(t)) \equiv A(t)$ can be obtained. The distribution functions f(X,t) and f(A,t) are related to each other by the equation

$$f(A,t) = \int \delta(A-A(X))f(X,t)dX, \quad \int f(A,t)dA = 1\,. \tag{1}$$

For this reason it is possible to define the <u>phase average</u> (the average with respect to the Gibbs ensemble) in two different ways ($\langle A(X)\rangle \equiv \bar{A}$)

$$\langle A(t)\rangle = \int Af(A,t)dA = \int A(X)f(X,t)dX\,. \tag{2}$$

Let us consider now the <u>time average</u> for the realization A(t) within the interval (t,t+T). It is defined by

$$\tilde{A}^T(t) = \frac{1}{T}\int_t^{t+T} A(t')dt'\,. \tag{3}$$

This equation can be written in the form (2)

$$\tilde{A}^T(t) = \int A\tilde{f}^T(A,t)dA \tag{4}$$

if the distribution function

$$\tilde{f}^T(A,t) = \frac{1}{T}\int_t^{t+T} \delta(A-A(t'))dt' \;,\quad \int \tilde{f}^T(A,t)dA = 1 \tag{5}$$

is introduced.

$\tilde{f}^T(A,t)dA$ gives the relative residence time of A(t) in the infinitesimal interval dA around the value A within the time interval (t,t+T).

From (4) and (5) the equivalence of the definitions (3) and (4) follows.

If during the time τ_{rel} the equilibrium or a stationary state is established in the system the distribution function is for

$$T \gg \tau_{rel} \tag{6}$$

in zeroth order with respect to the parameter τ_{rel}/T given by

$$\tilde{f}(A) = \lim_{T\to\infty} \tilde{f}^T(A,t) = \lim_{T\to\infty} \frac{1}{T}\int_t^{t+T} \delta(A-A(t'))dt' \quad . \tag{7}$$

This distribution is independent of t and coincides with the corresponding distribution over the Gibbs ensemble. Time and phase averages are the same, i.e.

$$\tilde{A} = \lim_{T\to\infty} \tilde{A}^T = \langle A\rangle \;. \tag{8}$$

This equation denotes the condition of ergodicity.

It is only valid in equilibrium or for steady states for which the limit $T \to \infty$ is possible. In this connection the question about the deviation of time averages from phase averages in the case of finite intervals of averaging is important.

Let us return to formula (5). In addition to the smoothed function $\tilde{f}^T(A,t)$ one can introduce a microscopic (unsmoothed) distribution function

$$f^M(A,t) = \delta(A-A(t)) \;,\quad \int f^M dA = 1 \;. \tag{9}$$

The corresponding average

$$\int Af^M(A,t)dA = A(t) \tag{10}$$

is, of course, identical with the function A(t) itself. Besides $\tilde{A}^T(t)$ there are other important statistical characteristics e.g. the effective volume of the realization A(t) in the space A within the interval (t,t+T). Generally, this volume depends on both the lengths and the position of the interval. For this reason we denote it by $\Delta\tilde{A}^{(T)}(t)$.

Depending on the specific problem, it is convenient for later calculations to choose either the value A(t) at the beginning of the interval (t,t+T) or the average $\tilde{A}^T(t)$ as the "centre" of the effective phase volume $\Delta\tilde{A}^{(T)}(t)$. In what follows we use the second possibility.

We define the effective phase volume as the minimal volume in phase space for which the distribution function $\tilde{f}^{(T)}(A,t)$ (cf. (5))

in the point $\tilde{A}^{(T)}(t)$ satisfies the equation

$$\int \tilde{f}^{(T)}(A,t)dA = \tilde{f}^{(T)}(\tilde{A}^T(t),t) \;\; \Delta\tilde{A}^{(T)}(t) = 1 \,. \tag{11}$$

This definition ensures that the volume $\Delta\tilde{A}^{(T)}(t)$ depends on t and T but is independent of the variable A. Another important stochastic characteristic is the mean distance between the trajectories within the interval (t,t+T). In order to define it in the form $A(t) \equiv A(X(t))$ we use the function X(t) and denote by D(t) and $D(t_o)$ the distance between two trajectories at time point t and t_o, respectively.

$$D(t) = \sqrt{[X_1(t)-X_2(t)]^{\,2}} \,, \quad D(t_o) = \sqrt{[X_1(t_o)-X_2(t_o)]^{\,2}} \,. \tag{12}$$

These quantities are microscopic (unsmoothed) characteristics. Besides them, functions are used which are smoothed with the help of two-trajectory distribution functions

$$\tilde{f}^{(T)}(X_1,X_2,t) = \frac{1}{T}\int_t^{t+T} \delta(X_1-X_1(t))\, \delta(X_2-X_2(t))dt \,,$$

$$\int \tilde{f}^{(T)} dX_1 dX_2 = 1 \,. \tag{13}$$

The smoothed distance between two trajectories is then given by

$$\tilde{D}^{(T)}(t) = \int \sqrt{(X_1-X_2)^2}\; \tilde{f}^{(T)}(X_1,X_2,t)dt. \tag{14}$$

Integrating over X_1, X_2 and using (13) it can be written in the form

$$\tilde{D}^{(T)}(t) = \frac{1}{T}\int_t^{t+T} \sqrt{[X_1(t') - X_2(t')]^2} \;\; dt' \,. \tag{15}$$

For the calculation of Ljapunov exponents instead of (12), (14) the quantities

$$D_i(t) = \sqrt{[X_{1i}(t) - X_{2i}(t)]^{\,2}} \,, \; i = 1,2,\ldots,N \tag{16}$$

are used, which are the distances with respect to the corresponding coordinates. N denotes the number of degrees of freedom. The concrete choice of the coordinate system is, of course, determined by the problem and its appropriateness for the necessary calculations.

2. Entropy in Time Averaging. Boltzmann Principle

The entropy corresponding to the smoothed function $\tilde{f}^{(T)}(A,t)$ is defined by

$$\tilde{S}^{(T)}(t) = - \int \ln \tilde{f}^{(T)}(A,t)\; \tilde{f}^{(T)}(A,t)dA + S_o \,. \tag{17}$$

We recast this expression with the help of (11) for the effective phase volume. To this end we replace in (17) the argument of the log $\tilde{f}^{(T)}(A,t) \longrightarrow \tilde{f}^{(T)}(\tilde{A}^{(T)},t)$ and use the definition (11) of the effective phase volume $\Delta\tilde{A}^T(t)$. We find

$$\tilde{S}^{(T)}(t) = \ln \Delta\tilde{A}^{(T)}(t) \int \tilde{f}^{(T)}(A,t)dA + S_o \,. \tag{18}$$

Using the normalization condition of the function $\tilde{f}^{(T)}$ we get the following expression for the entropy

$$\tilde{S}^{(T)}(t) = \ln \Delta \tilde{A}^{(T)}(t) + S_0 \ . \tag{19}$$

Equation (19) plays the role of Boltzmanns relation between entropy and logarithm of the phase volume or thermodynamic probability in the case of time averaging. From (19) we find the entropy difference for two time points t and t_0

$$\tilde{S}^{(T)}(t) - \tilde{S}^{(T)}(t_0) = \ln \frac{\Delta \tilde{A}^{(T)}(t)}{A^{(T)}(t_0)} \ . \tag{20}$$

Three cases can be distinguished: 1. The effective phase volume shrinks in the course of time, the entropy decreases; 2. the effective phase volume and the entropy remain constant; 3. the effective phase volume grows with time, the entropy increases. The first case is to be found in open systems only. Case two occurs for reversible processes taking place either in Hamiltonian systems (systems without dissipation) or in equilibrium. The third case occurs in closed systems which relax into the equilibrium state.

From (20) we find the relation

$$\Delta \tilde{A}^{(T)}(t) = \Delta \tilde{A}^{(T)}(t_0) \exp\{ \tilde{S}^{(T)}(t) - \tilde{S}^{(T)}(t_0)\} , \tag{21}$$

which expresses the Boltzmann principle in the case of time averaging.

3. Entropy Change in Time Averaging

Dividing both sides of (21) by $(t-t_0)$ we get a balance equation relating the mean variation of the entropy during the time $(t-t_0)$ to the corresponding entropy change.

$$\frac{\tilde{S}^{(T)}(t) - \tilde{S}^{(T)}(t_0)}{t - t_0} = \tilde{\sigma}^{(T)}(t,t_0) \equiv \frac{1}{t-t_0} \ln \frac{\Delta \tilde{A}^{(T)}(t)}{\Delta \tilde{A}^{(T)}(t_0)} \ . \tag{22}$$

Denoting the relaxation time to equilibrium or to a stationary state by τ_{rel} and assuming only small changes of the entropy production in the interval $t - t_0 \ll \tau_{rel}$, i.e.

$$\tilde{\sigma}^{(T)}(t,t_0) = \tilde{\sigma}^{(T)}(t) + 0 \left(\frac{t-t_0}{\tau_{rel}}\right) , \tag{23}$$

we can write (22) in the form of a balance equation

$$\frac{d\tilde{S}^{(T)}(t)}{dt} = \tilde{\sigma}^{(T)}(t) \ . \tag{24}$$

Depending on the way in which the phase volume changes in the course of evolution the entropy can show different time behaviour. In an isolated system the phase volume either grows or remains constant. Hence the inequality

$$\frac{d\tilde{S}^{(T)}}{dt} = \tilde{\sigma}^{(T)}(t) \geqslant 0 \tag{25}$$

is valid, which expresses the H-theorem in the case of time averaging.

In open systems the sign of the entropy change is not constrained. The H-theorem remains valid only under the additional condition that during the evolution to the stationary state the value of the entropy is renormalized in such a way, that the mean energy remains constant. This is the H-theorem for open systems in which stationary states exist.

Instead of the entropy referring to a constant value of the mean energy also other Ljapunov-functions can be used, e.g. a in a particular manner defined "free energy of the nonequilibrium state". Such a description is, however, less general and less natural. Another approach to the comparison of the degree of order of different states of an open system uses a sequence of stationary states corresponding to different values of the control parameter a. The values of the entropy obtained in this way are once more renormalized to the same value of the mean energy. Whenever these values of the (renormalized) entropy decrease, a process of selforganization takes place in the system. Given several control parameters, different paths of evolution become possible. On the basis of the S-theorem one can look for optimal paths of evolution for these systems.

4. The Krylov-Kolmogoroff Entropy

The Krylov-Kolmogoroff entropy is defined by the equation

$$K(t) = \frac{1}{t-t_0} \ln \frac{D(t)}{D(t_0)} . \tag{26}$$

The quantities $D(t)$ and $D(t_0)$ (cf. (12)) are microscopic characteristics which due to the existence of smoothing influence cannot be determined exactly in numerical or real experiments.

More realistic is the corresponding smoothed quantity

$$\tilde{K}^{(T)}(t) = \frac{1}{t-t_0} \ln \frac{\tilde{D}^{(T)}(t)}{\tilde{D}^{(T)}(t_0)} , \tag{27}$$

which is defined with the help of the smoothed functions (14,15). As is well known, the K-entropy is related to the Ljapunov exponents. The spectrum of the Ljapunov exponents characterizes the degree of disorder of the considered states of a nonlinear system. A positive Ljapunov exponent, e.g., indicates the existence of a strange attractor. The Ljapunov exponents are, however, determined by using linearized equations for $X(t)$ and $A(t)$. For this reason it remains an open question whether one can completely define the level of chaoticity by using only the spectrum of Ljapunov exponents. Similar problems arise for other well known criteria of the level of chaoticity as well. It is important that one can characterize the level of chaoticity on the basis of the S-theorem without relying upon the linearizations of the equations under consideration.

References

W. Ebeling, Yu.L. Klimontovich: Selforganization and Turbulence in Liquids (Teubner, Berlin 1984)

H. Haken: Synergetics (Springer, Berlin Heidelberg New York 1978; Mir, Moscow 1980)

H. Haken: Advanced Synergetics (Springer, Berlin Heidelberg New York Tokyo 1983; Mir, Moscow 1984)

Yu.L. Klimontovich: Statistical Physics (Nauka, Moscow 1982; Gordon and Breach, Harwood Academic Publishers, New York 1985)

Yu.L. Klimontovich: Brownian motion and turbulence: entropy, entropy production with laminar and turbulent motions. In: Nonlinear and Turbulent Processes in Physics (Gordon and Breach, Harwood Academic Publishers, New York 1984)
G. Nicolis, I. Prigogine: Selforganization in Nonequilibrium Systems (Wiley, New York 1977; Mir, Moscow 1979)
N.S. Krylov: Works on the Foundation of Statistical Physics (Nauka, Moscow 1950; Princeton University Press, Princeton 1979)
G. Nicolis, G. Dewel, J. Turner: Order and Fluctuations in Equilibrium and Nonequilibrium Statistical Mechanics, XVIIth International Solvay Conference on Physics (Wiley, New York 1981)
I. Prigogine, I. Stengers: Order out of Chaos (Bantam Books, Toronto New York London 1984)
Z.M. Zaslavskij: Stochasticity of dynamical systems (Nauka, Moscow 1984)
A.J. Lichtenberg, M.A. Lieberman: Regular and Stochastic Motion (Springer, Berlin Heidelberg New York 1983; Nauka, Moscow 1984)
Yu.L. Klimontovich: Turbulent Motion. The Structure of Chaos (Springer, in press)

Nucleation Theory and Dynamics of First-Order Phase Transitions in Finite System

H. Ulbricht, F. Schweitzer, and R. Mahnke

Wilhelm-Pieck-Universität Rostock, Sektion Physik, Universitätsplatz 3, DDR-2500 Rostock, GDR

1. Introduction

If we consider a vapor at equilibrium, then a certain change of the thermodynamic parameters (e.g. a supercooling at a given density, see Fig. 1) is able to remove the system into a nonequilibrium state. The vapor thus becomes supersaturated and a phase transition can occur in the system.

Inside the binodal region, the mechanism of such a first-order phase transition is the homogeneous nucleation, if impurities are absent. In general, we can divide the process of phase separation into three stages: nucleation, growth and coarsening. During the first stage, in the supersaturated vapor small nuclei are formed by thermal fluctuations. The critical size of these embryos is caused by the formation of their surface. Overcritical nuclei can grow to form large macroscopic domains (droplets, see also Fig. 2), undercritical nuclei diminish again.

If we take into account a depletion of the vapor in a finite system, that means the vapor pressure decreases because of nucleation, at a certain value of supersaturation no more overcritical droplets are able to arise and the growth process of the established droplets is dominating. For a more decreased vapor pressure the growth is converted to a competition process of the droplets, that is the so-called Ostwald ripening.

Despite intensive research over a period of about sixty years, a number of problems concerning the theoretical description of the

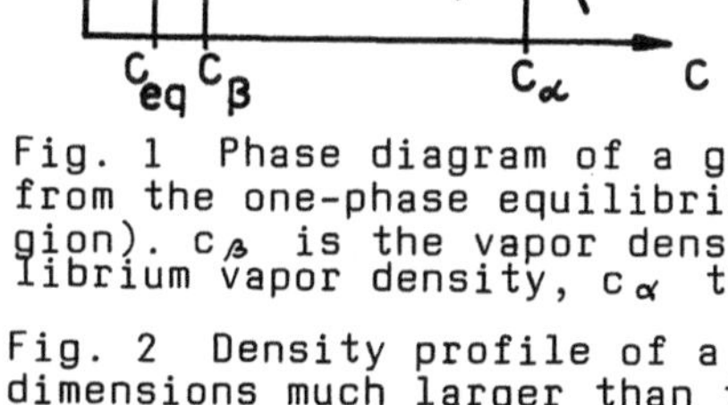

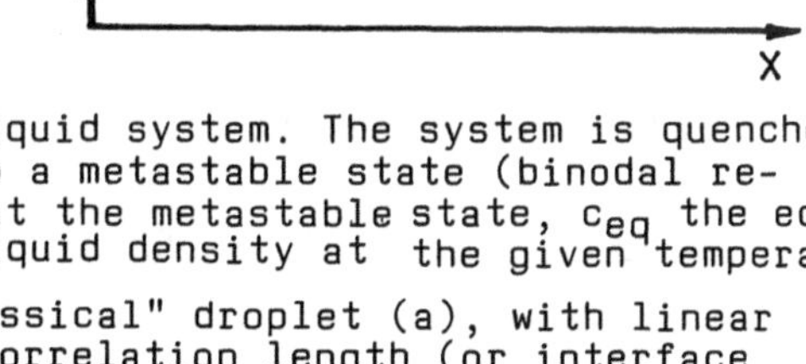

Fig. 1 Phase diagram of a gas-liquid system. The system is quenched from the one-phase equilibrium to a metastable state (binodal region). c_β is the vapor density at the metastable state, c_{eq} the equilibrium vapor density, c_α the liquid density at the given temperature

Fig. 2 Density profile of a "classical" droplet (a), with linear dimensions much larger than the correlation length (or interface thickness) d_o and of a "nonclassical" droplet (b) of smaller size, where interior and surface region are no longer clearly distinguishable /10/

nucleation process remain not finally solved yet. Some of these problems are the calculation of the formation energy of critical clusters, the proper consideration of the depletion of the initially metastable phase, the kinetic description of the relaxation into the stable state of the coexisting phases and the influence of the finiteness of the system.

Several models have been pointed out to clarify this situation (/1-3/, see also /4,5/). The classical nucleation theory mainly tends to calculate the stationary nucleation rate for the formation of overcritical droplets in an infinite system using a one-parametric description of the droplets. This theory was modified recently by several authors /6-10/ completing the droplet model. By means of a continuous description of the phase-separating system /11-14/ a more general level of the theory is reached that allows the determination of some kinetic properties near the critical point /15-17/. Recently, computer simulations have given a useful demonstration of the earlier stages of the nucleation process and of the time-dependent cluster evolution /18-20/. They also allow the calculation of the formation energy of microclusters /21/.

For the late stage of the phase transition, the asymptotic solution of the kinetic equations can be obtained from the theory of Ostwald ripening /22-24/. A generalization of this result is given by recent authors applying thermodynamic investigations /25,26/. A thermodynamic analysis can lead to a deeper insight into the process of phase separation, in particular, it gives general results for the existence of stable droplets in principle and their dependence on the thermodynamic parameters of the system, like particle number, system size, pressure and temperature /27-29/.

The classical and other nucleation theories suppose an infinite system neglecting the alteration of the system properties caused by the nucleation process. In finite systems, on the contrary, the limited system size and/or particle number yields a rather complicated evolution of the droplet distribution (see also /30/). Another important contribution to a better description of the kinetic and the stationary properties is that of a proper theory of surface effects, because the surface tension has a great influence on the formation of critical droplets. While classically the capillarity approximation is used, modern theories apply curvature-depending surface tensions resulting from statistic /21,31/ or thermodynamic /32,33/ investigations.

A complete theory of the dynamics of first-order phase transitions, including the stochastic formation of nuclei as well as the late stage competition process within the droplet ensemble, would be very helpful for a variety of physical processes. Many areas of current interest in the material sciences involve the simultaneous diffusional growth and dissolution of domains in a matrix /34/.

In this paper we describe a heterogeneous system with a finite volume consisting of monomers (free particles) and different drops (n-mers) as bound states. After studying the stochastic and kinetic evolution of one droplet, we find for an ensemble of drops under constrained conditions that larger drops grow at the expense of smaller drops, which disappear. So the Ostwald ripening as a competition process quite similar to the selection in bio- or eco-systems is related to the theory of selforganization /35/.

2. Stochastic Evolution of One Droplet in a Finite System

In the following we consider the stochastic evolution of a droplet in a supersaturated vapor. The thermodynamic constraints of the system we fix as follows:

$$N = \text{const}, \quad V = \text{const}, \quad T = \text{const} \,, \tag{2.1}$$

where N is the overall particle number, V the volume and T the temperature of the system. The thermodynamic parameters must be chosen in such a way, that the pressure for the (supposed) ideal vapor

$$p = \frac{N}{V} k_B T \tag{2.2}$$

is much larger than the equilibrium pressure p_∞ of the saturated vapor. If we define an initial supersaturation

$$y_o = \frac{p - p_\infty}{p_\infty} , \tag{2.3}$$

then $y_o > y_{cr}$, where y_{cr} is a critical supersaturation as pointed out in chapter 3, gives one permission that a nucleation process can occur. If we take into account the alteration of the pressure due to nucleation, the actual supersaturation y(t) decreases with time, starting from y_o.

We suppose that a single droplet was formed in the system and evolves due to the following kinetic mechanism:

$$A_1 + A_l \underset{w^-}{\overset{w^+}{\rightleftharpoons}} A_{l+1}$$

$$(N-l)A_1 \underset{w^-}{\overset{w^+}{\rightleftharpoons}} (N-l-1)A_1 , \tag{2.4}$$

where l is the number of particles bound in the droplet ($l \leq N$). Both reactions (2.4) are coupled by the limitation of the overall particle number.

The evolution of the droplet is assumed to be a Markovian birth and death process. If we define P(l,t) as the probability to find the droplet with l particles at a given time t, the following master equation describes the time dependence of P(l,t) /36/:

$$\dot{P}(l,t) = w^+(l-1)P(l-1,t)+w^-(l+1)P(l+1,t)-P(l,t)[w^+(l)+w^-(l)] . \tag{2.5}$$

w^+ and w^- are the transition probabilities per unit time as also introduced in (2.4). We define them in agreement with former investigations /30,37/ as follows:

$$w(N-l-1,l+1|N-l,l) \equiv w^+(l) = \alpha l^{2/3} \frac{N-l}{V}$$

$$w(N-l+1,l-1|N-l,l) \equiv w^-(l) = \alpha l^{2/3} \frac{1}{\lambda_1^3} \exp\left(\frac{f_l - f_{l-1}}{kT}\right), \quad 2 \leq l \leq N \tag{2.6}$$

$\lambda_1 = h(2\pi m k_B T)^{-1}$ is the de Broglie wave length of a free particle with the mass m and $f_l = -Al + Bl^{2/3}$ is a potential function with $f_1 = 0$. A and B are constants, given by /38/:

$$A = - \ln \frac{p_\infty}{k_B T} \lambda_1^3 \quad , \quad B = 4\pi\left(\frac{4\pi}{3} c_\alpha\right)^{-2/3} \sigma . \tag{2.7}$$

α is a constant with respect to the special properties of the droplet, like the liquid density c_α [particles/m³], the surface tension σ [N/m] and the temperature T [K] .

Note, that in the case l=1 it holds /30/:

$$w^+(1) = \alpha \frac{N(N-1)}{V} \quad ; \quad w^-(1) = 0 \quad . \tag{2.8}$$

The transition probabilities for processes different from (2.4) are set equal to zero.

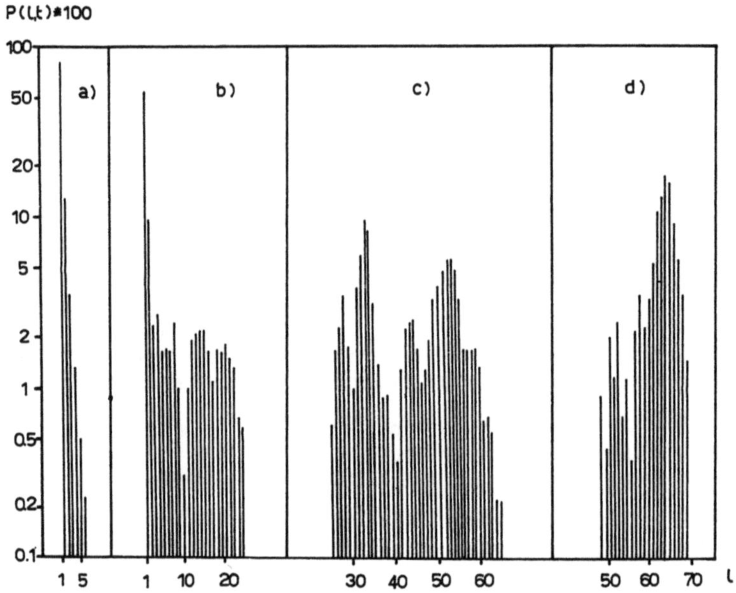

Fig. 3 Time evolution of a vapor-liquid phase transition in a finite system. $P(l,t)$ is the probability of the droplet to be in state l. Four different evolution stages, averaged over 500 elementary processes, are shown corresponding to the (reduced) time a) 7500; b) 16000; c) 29500; d) 33000.
vapor: ethanol, $T = 290$ K, $N = 150$, $V = 2{,}75 \cdot 10^{-23}$ m^3

A computer simulation with the given transition probabilities is demonstrated in Fig. 3 for several stages of the evolution of the droplets.

We can separate two characteristic time scales of the process (see also /40-42/). In a short time regime a quasistationary distribution for the vapor phase is established around $l=1$, while in a long time regime a distribution around the second stable state evolves by means of the non-poissonian fluctuations. The development of this stable state just corresponds to the phase transition, because the second maximum of $P(l,t)$ is due to the droplet phase. Transition between the two stable states are possible and will be discussed in the following.

3. Equilibrium Distribution and Bistability

The master equation (2.5) allows us to introduce a probability flux

$$J(l-1,l) = w^+(l-1)P(l-1,t) - w^-(l)P(l,t) \quad , \tag{3.1}$$

which is analogous to the classical nucleation rate. In the stationary case it holds:

$$J(l-1,l) = J(l+1,l) = J^s = \text{const} \tag{3.2}$$

With this condition we receive the stationary probability distribution in a way given by BECKER and DÖRING /1/ (see also ref. /6/):

$$P^s(m) = P^s(l) \prod_{j=l+2}^{m} \frac{w^+(j-1)}{w(j)} - J^s\Big(\sum_{k=l}^{m-1} Q_k\Big) \ , \quad m \geqslant l \quad \text{with} \tag{3.3}$$

$$Q_k = \frac{1}{w^+(k)} \prod_{j=l+1}^{m} \frac{w^+(j-1)}{w^-(j)} \quad .$$

For $P^s(m) = 0$, that means, m is an absorber state, we get with l=1 the stationary probability flux

$$J^s = P^s(1)\left\{\sum_{k=1}^{m-1} \frac{1}{w^+(k)} \frac{1}{\prod_{j=2}^{k} \frac{w^+(j-1)}{w^-(j)}}\right\}^{-1} \quad . \tag{3.4}$$

The equilibrium distribution $p^o(l)$ we obtain from (2.5) with the condition of detailed balance /36/:

$$w^+(l)P(l,t) = w^-(l+1)P(l+1,t) \tag{3.5}$$

that means consequently $J^s = 0$. From (3.3) it yields:

$$P^o(l) = P^o_N(1) \prod_{j=2}^{l} \frac{w^+(j-1)}{w^-(j)} \tag{3.6}$$

with the normalization

$$P^o_N(1) = \left\{1 + \sum_{n=2}^{N} \prod_{j=2}^{n} \frac{w^+(j-1)}{w^-(j)}\right\} .$$

Figure 4 shows that the equilibrium distribution $p^o(l)$ is able to be bimodal in dependence on the thermodynamic parameters.

The existence of two stable states of $P^o(l)$ represents the coexistence of the liquid and the vapor phase in equilibrium. The second maximum of $P^o(l)$ defines the stable droplet size l_{st}, while the minimum of $P^o(l)$ (not to be seen in Fig. 4) is related to the

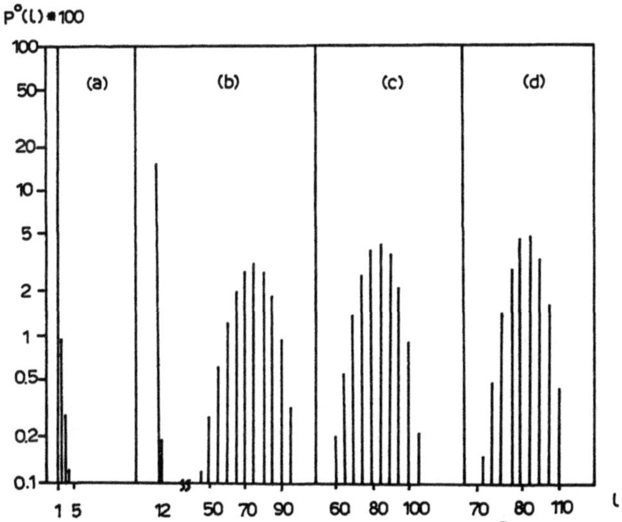

Fig. 4 Equilibrium distribution $P^o(l)$ for various values of the system volume V. The maximum states are relative to the vapor (l=1) and the droplet ($l=l_{st}$). In dependence of the system size we find a coexistence of the droplet and the vapor
vapor: ethanol, T = 290 K, N = 150

a) $V = 2.75\cdot10^{-23}m^3$; b) $V = 2.5\cdot10^{-23}m^3$; c) $2.25\cdot10^{-23}m^3$

d) $V = 2.00\cdot10^{-23}m^3$ /38/

critical droplet size l_{cr}. Both values l_{st} and l_{cr} strongly depend on the values of the thermodynamic parameters /29,39/.

From a stochastic point of view, transitions between the two stable states are possible if $P^o(l)$ is bimodal. The mean first passage time for the transition between two states l and m (m ≥ l) can be calculated as follows:

$$\tau(l \rightarrow m) = \sum_{j=l}^{m-1} \frac{1}{w^+(j)} \frac{1}{P^o(j)} \sum_{k=1}^{j} P^o(k) . \tag{3.7}$$

In Tab. 1 there is demonstrated the dependence of the mean first passage time on the initial supersaturation y_o (2.3), which caused the distinction of the second stable state of $P^o(l)$

Tab. 1 Mean first passage time to reach the critical state l_{cr} and the stable state l_{st} in dependence on the system volume V and supersaturation y_o /39/

V ($10^{-23}m^3$)	y_o	l_{cr}	l_{st}	$\tau(1 \rightarrow l_{cr})$	$\tau(1 \rightarrow l_{st})$
2.0	4.79	9	93	372.7	830.5
2.25	4.15	12	85	1607.3	3333.1
2.5	3.63	15	75	7237.6	16616.4
2.75	3.21	21	63	51761.4	119631.6

Further we want to discuss the problem under which thermodynamic constraints the bimodality of $P^o(l)$ is just diminishing. That means a stable droplet is impossible to exist then and we find only a one-phase equilibrium state.

From the extremum condition for the equilibrium distribution

$$P^o(l^E) = P^o(l^E + 1) , \tag{3.8}$$

we get with the condition of detailed balance (3.5) and with the introduced transition probabilities (2.6) the relation:

$$\ln\left(\frac{l}{l+1}\right)^{2/3} + \ln \frac{N-l}{V} \lambda_1^3 = \frac{1}{k_B T} \left\{ f_{l+1} - f_l \right\} . \tag{3.9}$$

For the extremum states of $P^o(l)$ it follows approximately:

$$\ln \frac{(N-l)\, k_B T}{p_\infty V} = \left(\frac{2\sigma}{c_\alpha k_B T}\right)\left(\frac{4\pi}{3} c_\alpha\right)^{1/3} l^{-1/3} . \tag{3.10}$$

A further discussion of (3.10) /38/ leads to the condition (3.11) that gives a relation between the critical values of the thermodynamic parameters N, V and T for which the bimodality of $P^o(l)$ is lost:

$$N \left\{ \ln \frac{N k_B T}{p_\infty V} \right\}^4 = 4 \left(\frac{4\pi}{3} c_\alpha\right) \left\{ \frac{4}{3} \frac{2\sigma}{c_\alpha k_B T} \right\}^3 . \tag{3.11}$$

Note, that the equilibrium pressure p_∞, the surface tension σ and the particle density c_α of the liquid depend on the temperature too.

From (3.11) results a critical overall particle number N_c respectively a critical system size V_c as well as a critical temperature T_c of the system, where a stable coexistence of a droplet in the vapor is just impossible. This fact is a consequence of the finiteness

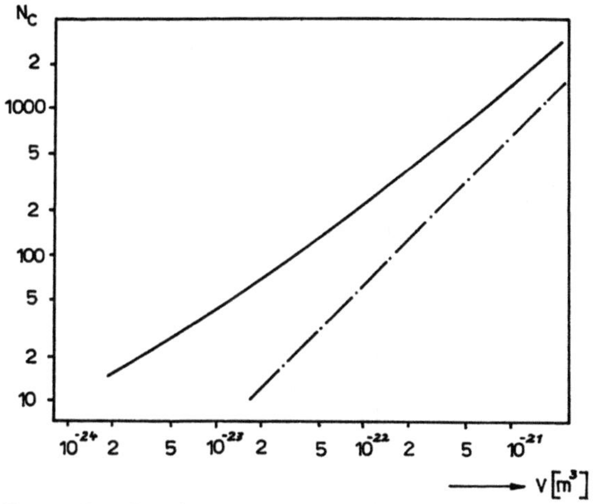

Fig. 5 Critical overall particle number N_c vs. system volume V. Only for $N > N_c$ is a coexistence of the droplet in the vapor possible. The dashed-dotted line gives the saturation particle number $N_\infty = p_\infty V/k_B T$ /38/
vapor: ethanol, T = 280 K

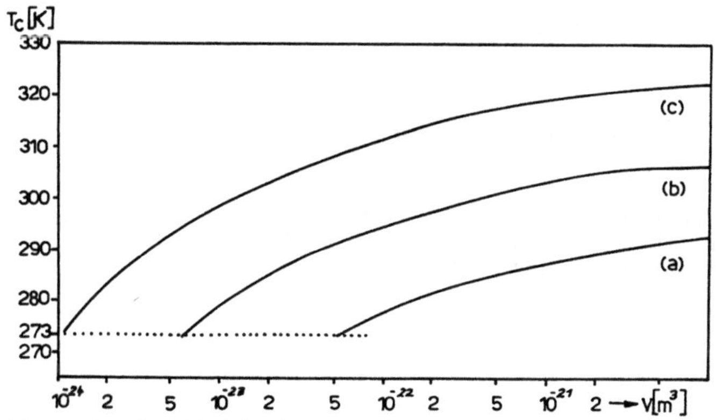

Fig. 6 Critical temperature T_c vs. system volume V. If $T > T_c$, the binodal region of the finite system is left and no nucleation can be obtained. T_c is plotted for various overall particle densities:
a) $N/V = 2\ 10^{24} m^{-3}$; b) $N/V = 4\ 10^{24} m^{-3}$; c) $N/V = 8\ 10^{24} m^{-3}$ /38/
vapor: ethanol

of the system. The region where a two-phase coexistence of the finite system is possible is shown in Fig. 5 and Fig. 6. We want to underline, that the critical temperature of the binodal region depends on the system size too, this fact should become important for small systems.

The existence of critical thermodynamic parameters for the nucleation process permits the calculation of a critical supersaturation of the finite system /38/

$$y_c = \frac{p_c - p_\infty}{p_\infty} = \exp\left\{ \left(\frac{\frac{4\pi}{3} c_\alpha}{\frac{N_c}{4}} \right)^{1/4} \left(\frac{4}{3} \frac{2\sigma}{c_\alpha k_B T} \right)^{3/4} \right\} - 1 . \qquad (3.12)$$

For the initial supersaturation $y < y_c$ the system possesses only one stable state due to the vapor phase. An overcritical droplet is not able to be formed because the pressure of the system decreases faster due to the attachment of free particles by the droplet, than the droplet reaches the overcritical size. This effect is caused by the finiteness of the system.

For $y > y_c$ a stable droplet is possible to exist. That means y_c has to be reached at least initially to insert the nucleation process. y_c gives the so called cloud point for the phase transition in the finite system.

The value of y_c are given in Fig. 7.

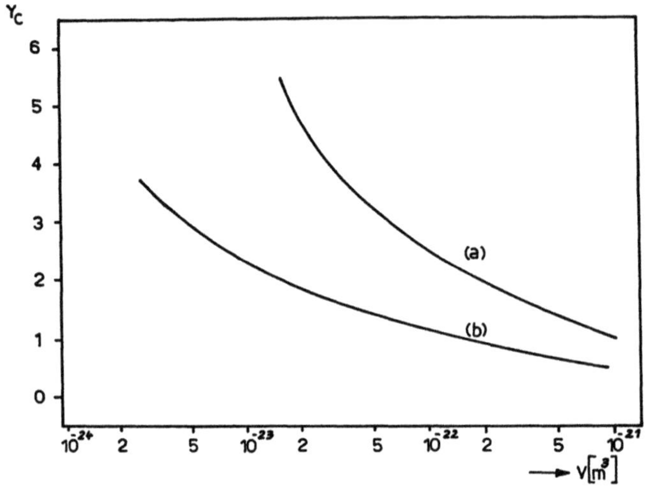

Fig. 7 Critical supersaturation y_c vs. system volume V. y_c gives the onset of nucleation, the so-called cloud point.
For $y < y_c$ a critical droplet is not able to be established.
a) T = 280 K; b) T = 312.15 K /38/
vapor: ethanol

4. Basic Equations of the Competitive Growth Problem

Apart from the concept discussed above, the time evolution of the phase transition can be described by means of a concentration field $c(\vec{r},t)$, which has to fulfill a reaction-diffusion equation or a similar functional master equation /44/. To avoid the difficulties in solving these equations, we consider a model where spherical droplets are located in the otherwise homogeneous vapor. Figure 8 (compare Fig. 2) indicates that our assumption of the droplet with a constant density c_α is no longer valid for small droplets. If the droplet radius r has the magnitude of the interface thickness d_o (explicitly if $r \leq d_o/2$) the droplet model breaks down, because there are no incompressible bulk regions. Then one has to take into account a changing density c_α like in Montroll's approach /45/. For a typical liquid-vapor system the capillary length is

$$d_o = 2\sigma (c_\alpha k_B T)^{-1} \sim 10^{-9} m = 1 \text{ nm} . \qquad (4.1)$$

In the following we use macroscopic values for the constants c_α (density of monomers within the droplet) $c_{eq}(\infty)$ (equilibrium concentration of monomers), σ (surfce tension of the droplets), T

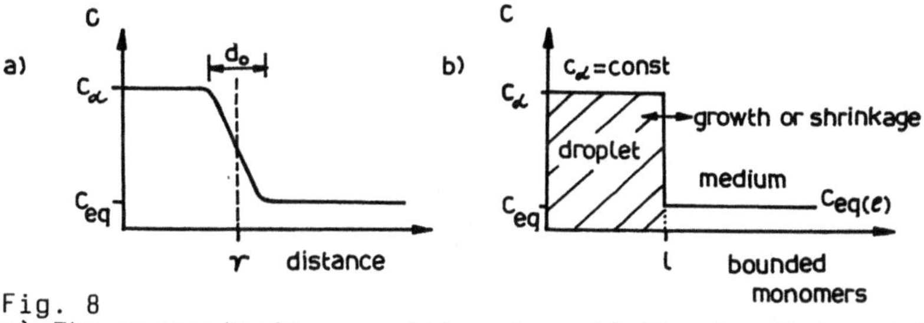

Fig. 8

a) The concentration profile of the moving droplet interface and the definition of the droplet radius r(t)

b) The droplet model with an incompressible region of bound monomers and free monomers in the medium

(temperature), D (diffusion coefficient of monomers) coming from experimental data. Now our basic quantities describing the system are

$$c(t) = M_\beta(t)/V \qquad \text{concentration of free monomers} \qquad (4.2)$$

$$f(l,t) = N(l,t)/V \qquad \text{cluster density distribution function} \qquad (4.3)$$

where $M_\beta(t)$ is the number of the free monomers in the volume of the system V and N(l,t) the number of droplets with l bounded particles respectively. The time-dependent supersaturation characterizing the monomer phase thus can be written in the form

$$y(t) = (c(t) - c_{eq}(\infty))/c_{eq}(\infty) \geq 0 \;, \qquad (4.4)$$

where $c_{eq}(\infty)$ is the concentration of monomers in equilibrium with an infinitely large droplet (planar surface) : $c_{eq}(\infty) = p_{eq}/k_B T$.

To establish appropriate scales, we may introduce a "characteristic unit of time"

$$t_o = c_\alpha d_o^2 (c_{eq}(\infty)D)^{-1} \sim 10^{-9}\text{s} = 1 \text{ ns} \qquad (4.5)$$

for defining a reduced time $\tau = t/t_o$ instead of the dimensional time t, and a "characteristic unit of system size"

$$V_o = (4\pi/3)d_o^3 c_\alpha / c_{eq}(\infty) \approx 10^{-23}\text{m}^3 = 10^4\text{nm}^3 \qquad (4.6)$$

to make use of a reduced volume $\omega = V/V_o$ instead of the system size V.

We choose the number l of bounded monomers as the relevant variable for a droplet, but the other quantities describing an l-mer can be easily defined as follows:

$$\begin{aligned} r(l) &= (c_\alpha 4\pi/3)^{-1/3} l^{1/3} && \text{radius} \\ k(l) &= (c_\alpha 4\pi/3)^{1/3} l^{-1/3} = 1/r(l) && \text{curvature} \\ V(l) &= c_\alpha^{-1} l && \text{volume} \\ A(l) &= 4\pi (c_\alpha 4\pi/3)^{-2/3} l^{2/3} && \text{surface} \end{aligned} \qquad (4.7)$$

We mention, that the distribution function (4.3) is normalized to the total droplet density (N_α - number of all drops)

$$\int_0^\infty f(l,t)dl = N_\alpha / V$$

and for the first moment of the distribution function we get the total number of bounded monomers M_α per volume

$$\int_0^\infty l f(l,t) dl = M_\alpha (t)/V \quad . \tag{4.8}$$

In the following, we formulate the basic equations of the competitive growth problem. The droplet distribution function (4.3), which changes due to condensation and/or evaporation of monomers only, satisfies a continuity equation of the form

$$\frac{\partial f(l,t)}{\partial t} + \frac{\partial}{\partial l}(v f(l,t)) = j(t) \quad , \tag{4.9}$$

with a growth rate of a droplet of size l in the presence of supersaturation y(t)

$$dl/dt = V(l,y) \quad . \tag{4.10}$$

The production-decay term j in (4.9) vanishes when nucleation or such phenomena like coagulation, splitting, etc., which introduce new droplets of a given size class, are neglectable.

Conservation of matter implies that for a closed system the total number N of monomers (either free or bound) is fixed

$$c(t) + \int_0^\infty l f(l,t) dl = N/V \quad ; \tag{4.11}$$

or with (4.4)

$$y(t) + (c_{eq}(\infty))^{-1} \int_0^\infty l f(l,t) dl = y_0 , \tag{4.12}$$

where $y_0 = (N/V - c_{eq}(\infty))/c_{eq}(\infty)$ = const is the overall supersaturation. In a recent paper of VENZL the parameter y_0 is time dependent for treating open systems /46/.

It is well known that growing and dissolving drops obey kinetic laws resulting from solutions of the diffusion equation. Employing the quasistationary approximation we assume that an individual drop grows (or dissolves) if the monomer concentration is greater (or less) than the equilibrium value. Therefore the growth law (4.10) which embodies much of the physics reads in the so-called interface kinetic limited case

$$dl/dt = v_{ik}(l,y) = D\, d_0^{-1} A(l)(c(t) - c_{eq}(l)) , \tag{4.13}$$

where the equilibrium monomer concentration $c_{eq}(l)$ over a curved surface with curvature k(l) is

$$c_{eq}(l) = c_{eq}(\infty) \exp(d_0 k(l)) \approx c_{eq}(\infty)(1 + d_0 k(l)) \quad . \tag{4.14}$$

This curvature-dependent expression is the GIBBS-THOMSON equation indicating that atoms will flow from regions of high to low curvature.

From (4.13), (4.14) we obtain

$$dl/dt = D c_{eq}(\infty) d_0^{-1} A(l)(y(t) - d_0 k(l)) \tag{4.15}$$

and after introducing the critical droplet size l_c by

$$l_c(t) = (c_\alpha\, 4\pi/3)(d_0/y(t))^3 = c_{eq}(\infty) v_0 y^{-3} \tag{4.16}$$

we are able to reformulate the kinetic law (4.10) in the form of the well-known droplet equation

$$dl/dt = Dc^{eq}(\infty)(c_\alpha 4\pi/3)^{1/3}A(l)(l_c^{-1/3} - l^{-1/3}) , \quad (4.17)$$

which makes apparent the typical bistability situation with opposite behaviour of drops below and above the critical value l_c. We want to underline, that due to the conservation of monomers (4.11) the critical droplet size is changing in time. Let us finally introduce a new measure of the droplet size in units of l_c by the dimensionless variable

$$z = l/l_c . \quad (4.18)$$

Therefore the basic equations (4.9-11) written in dimensionless variables (4.4), (4.18) and reduced time τ are

$$\frac{\partial F(z,\tau)}{\partial \tau} + \frac{\partial}{\partial z}(v(z,y)F(z,\tau)) = 0 , \quad (4.19)$$

$$y(\tau) + \frac{V_0}{y(\tau)^3}\int_0^\infty zF(z,\tau)dz = y_0 , \quad (4.20)$$

$$dz/d\tau = v(z,y,\dot{y}) = 3y^2z^{2/3}(1-z^{-1/3})+3y^{-1}z(dy/d\tau) , \quad (4.21)$$

with the scaled cluster distribution $F(z,\tau) = l_c f(l,\tau)$ instead of (4.3). This coupled nonlinear integro-differential system can serve as the starting point for the analysis of the competitive growth problem of different droplets /35/.

5. Growth of one Droplet in a Closed System

Here we want to investigate the deterministic behaviour of one single droplet in a changing medium which agrees with the stochastic description (chapter 2,3). From the basic equations (4.19-21) we get in this simple case

$$dz/d\tau = 3y^2z^{2/3}(1-z^{-1/3})+3y^{-1}z(dy/d\tau) , \quad (5.1)$$

$$y + z/(\omega y^3) = y_0 , \quad (5.2)$$

showing the kinetics $\dot{z} = v(z,y,\dot{y})$ (5.1) and the conservation of monomers (5.2). Recalling the parameter $\omega = V/V_0$ (4.6) as reduced system size we reformulate (5.1), (5.2) in the following form

$$dz/d\tau = (3/\omega)y^{-2}z^{2/3}(1-z^{-1/3})(\omega y^4-3z) \quad (5.3)$$

$$dy/d\tau = -(3/\omega)y^{-1}z^{2/3}(1-z^{-1/3}) \quad (5.4)$$

and with (5.2) $z = \omega y^3(y_0-y)$ we find

$$dy/d\tau = 3\omega^{-2/3}(y_0-y)^{1/3}-3\omega^{-1/3}(y_0-y)^{2/3} . \quad (5.5)$$

The analysis of the development of one droplet (5.3), (5.4) or (5.5) in dependence of the volume ω is shown in Fig. 9 and Fig. 10.

For fixed overall supersaturation $y_0 = 2$ we find a critical volume $\omega_c = 4^4/(3^3y_0^4) = 16/27 = 0{,}5926$ (or $V_c = 3{,}1\cdot10^{-23}m^3$). For very small systems ($\omega < \omega_c$) we observe free monomers only, but for $\omega > \omega_c$ the coexistence of droplets and monomers is obvious. Overcritical droplets grow to their stable size.

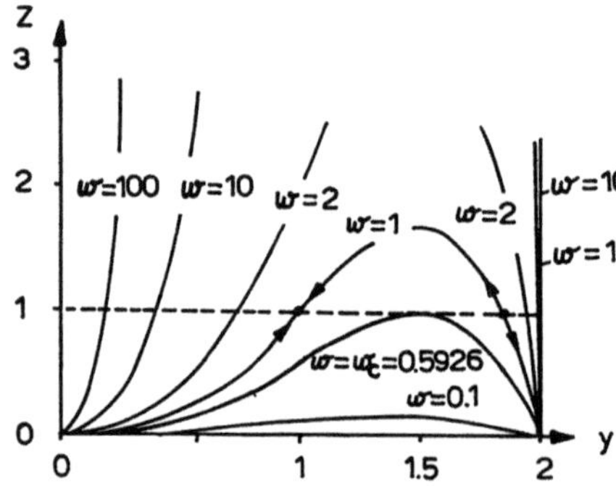

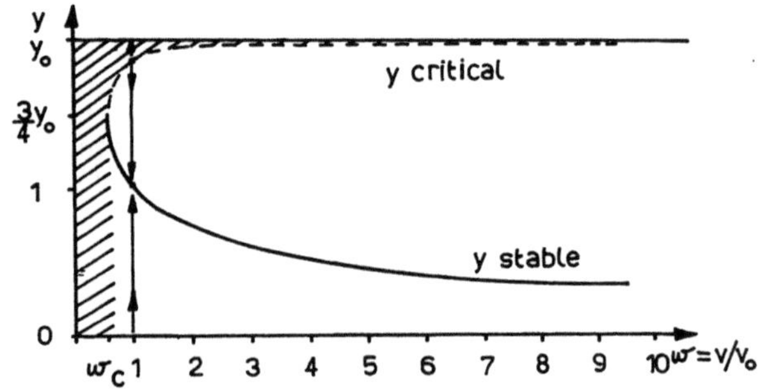

Fig. 9 Trajectory of droplet evolution in z-y-space in dependence on the dimensionless volume ω

Fig. 10 Plot of supersaturation y vs reduced volume ω showing the coexistence of a droplet and free monomers for $\omega \geq \omega_c$ and the area of monostability

6. Competitive Growth of an Ensemble of Droplets

To find solutions of the basic equations formulated in chapter 4 seems to be impossible. However, if the droplet ensemble consists of s kinds of drops with different sizes $l_1,\ldots,l_i,\ldots,l_s$, let us consider the discrete case. So we are able to obtain a system of ordinary nonlinear differential equations governing the constrained growth of droplets with $s \geq 1$ components /35/. Note that the index i is now the number of a component; we get a system of equations (6.1,6.2) which can be solved numerically

$$\frac{dl_i}{d\tau} = c_\alpha d_o A(l_i)\left\{ d_o(\langle k\rangle - k(l_i)) - \frac{c_{eq}(\infty)V}{c_\alpha d_o}\frac{1}{A_\alpha}\frac{dy}{d\tau}\right\} \qquad (i = 1,2,\ldots,s) \tag{6.1}$$

with

$$\frac{dy}{d\tau} = -\frac{c_\alpha d_o}{c_{eq}(\infty)V} A_\alpha(\tau)\left\{ y_o - \frac{c_\alpha}{c_{eq}(\infty)V} V_\alpha(\tau) - d_o\langle k\rangle\right\} \quad . \tag{6.2}$$

Here we abbreviate as follows: $A_\alpha(\tau)$ - total surface of all droplets, $V_\alpha(\tau)$ - total volume of all droplets, $\langle k\rangle$ - mean curvature of the droplet ensemble.

The equations describe the rapid growth (second term on r.h.s. of (6.1)) and the slow selection process (first term on r.h.s. of (6.1)). If the supersaturation is high enough, the droplets have the chance to grow up. Since the raw material is limited the system will reach at first a so-called internal equilibrium. When the flow between the liquid phase and the vapor is decreasing to zero, a competitive ripening process takes place. In this selection game the smaller droplets with $k(l_i) > \langle k\rangle$ must dissolve to give monomers to the bigger ones which increase. Since the mean curvature $\langle k\rangle(\tau)$ is a slowly decreasing function of time in the limit $\tau \to \infty$ only one component of the droplet phase is present and the selforganizing system has reached its stable stationary situation (Fig. 11).

It is obvious that our resulting equations (6.1,6.2) are a realistic physically motivated example for natural selforganization. It was pointed out by EIGEN /47/ and others /48/ that Darwinian evolution can be characterized by an extremum principle, which defines the behaviour of selfreplicative units. Under stated selec-

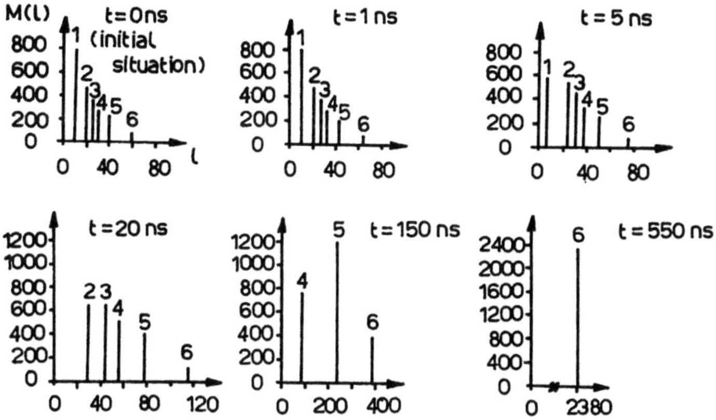

Fig. 11 Time evolution of a droplet ensemble showing the monomer distribution function $M(l,t) = l\cdot N(l,t)$. Numerical solution of (6.1,6.2) with s=6 groups of drops

tion constraints in bio- or ecosystems (compare the constant overall organization) the population numbers of all but one species will disappear. In the Fisher-Eigen model of prebiotical evolution

$$\frac{dx_i}{dt} = (E_i - \langle E \rangle)\, x_i \qquad \text{with} \quad \sum_{i=1}^{s} x_i = c \tag{6.3}$$

the species with the highest reproduction rate $E_m = \mathrm{Max}\{E_1,\ldots,E_s\}$ (selection value) will increase to the finite value C and all others must die out ($x \to 0$ for $i \neq m$). Note that in both cases (Fisher-Eigen model with population average fitness $\langle E \rangle$, Ostwald ripening of droplets with mean curvature of the droplet ensemble $\langle k \rangle$) the interaction of the different species is modeled only by an overall dilution flux which corresponds to the mean-field concept of many-body physics /49/. As we demonstrated, there is an interesting analogy in modeling phase transitions on the one hand and evolution processes in biophysics on the other.

References

1. R. Becker, W. Döring: Ann.Phys. 24, 719 (1935)
2. M. Volmer: Kinetik der Phasenbildung (Steinkopf, Dresden 1939)
3. J.L. Frenkel: Kinetic Theory of Liquids (Dover, New York 1955)
4. A.C. Zettlemoyer (Ed.): Nucleation (Dekker, New York 1969)
5. F.F, Abraham: Homogeneous Nucleation Theory (Academic Press, New York 1974)
6. J.L. Katz, M.D. Donohue: A kinetic approach to homogeneous nucleation theory, in: Adv.Chem.Phys. 40, 137 (1979); H. Wiedersich, J.L. Katz: Adv.Coll.Interface Sci. 10, 33 (1979)
7. J. Lothe, G.M. Pound: J.Chem.Phys. 36, 2080 (1962)
8. M.E. Fisher: Physics 3, 255 (1967)
9. B.E. Sundquist, R.A. Oriani: J.Chem.Phys. 36, 2604 (1962); 38, 2082 (1963)
10. K. Binder, D. Stauffer: Adv.Phys. 25, 343 (1976)
11. J.W. Cahn, J.E. Hilliard: J.Chem.Phys. 28, 258 (1958); 30, 1121 (1959); 31, 688 (1959)
12. J.S. Langer: Kinetics of metastable states, in: Lecture Notes in Physics 132 (Springer 1980)
13. K. Binder: Spinodal decomposition: in: Lecture Notes in Physics 132 (Springer 1980)

14. A.S. Patashinsky, B.I. Shumilo: JETP 77, 1417 (1979); FTT 22, 2609 (1980)
15. K. Binder: J.Physique C4, 51 (1980)
16. J.S. Langer, A.J. Schwartz: Phys.Rev. A29, 21 (1980)
17. H. Furukawa: Phys.Rev. A29, 2160 (1984)
18. O. Penrose, J. Lebowitz, J. Marro, M. Kalos, J. Tobochnik: J.Stat.Phys. 34, 399 (1983)
19. M. Rao, B.J. Berne, M.H. Kalos: J.Chem.Phys. 68, 1325 (1978)
20. W.H. Zurek, W.C. Schieve: J.Chem.Phys. 84, 1479 (1980)
21. F.F. Abraham: Rep.Progr.Phys. 45, 1113 (1982)
22. I.M. Lifschitz, V.V. Slyozov: Phys.Chem.Solids 19, 35 (1961)
23. C. Wagner: Z.Elektrochem. 65, 581 (1961)
24. J.A. Marqusee, J. Ross: J.Chem.Phys. 79, 373 (1983); J.A. Marqusee: J.Chem.Phys. 81, 976 (1984)
25. I. Gutzow, J. Schmelzer et al.: Rostocker Phys. Manuskr. 8, 22 (1985)
26. J. Schmelzer, F. Schweitzer: J.Non-Equilibr.Thermodyn., in press, Ann.Phys. (Leipzig), in press
27. W. Vogelsberger: Z.Phys.Chem. 258, 763 (1977); J.Coll.Interface Sci 88, 17 (1982)
28. J. Schmelzer, F. Schweitzer: Z.Phys.Chem. 266, 943 (1985)
29. F. Schweitzer, J. Schmelzer, H. Ulbricht: Wiss.Z. Rostock 33, 45, 54 (1984)
30. L. Schimansky-Geier, F. Schweitzer, W. Ebeling, H. Ulbricht: On the kinetics of isochoric gases, in this book
31. J.S. Rowlinson, B. Widom: Molecular Theory of Capillarity (Clarendon Press, Oxford 1984)
32. W. Vogelsberger et al.: Z.Phys.Chem. (Leipzig) 257, 580 (1976); 264, 625 (1983)
33. R. Mahnke, J. Schmelzer: Z.Phys.Chem. (Leipzig) 266, 1028 (1985); J. Schmelzer, R. Mahnke: J.Chem.Soc. Faraday Trans. 1 82, (1986)
34. P.W. Voorhees: J.Stat.Phys. 38, 231 (1985); P.W. Voorhees, M.E. Glicksman: Acta Metall. 32, 2001, 2013 (1984)
35. R. Mahnke, R. Feistel: in: Nonlinear Irreversible Processes and Phase Transitions, Rostocker Phys. Manuskr. 8, 54 (1985); R. Mahnke: Wissenschaftl. Zeitschrift Wilhelm-Pieck-Universität Rostock, N-Reihe 35, (1986)
36. H. Haken: Synergetics - An Introduction (Springer, Berlin Heidelberg New York 1978)
37. W. Ebeling: Sitzungsberichte der AdW der DDR 22N, 33 (1981)
38. F. Schweitzer, L. Schimansky-Geier: submitted for publication
39. F. Schweitzer, A. Budde, H. Ulbricht: submitted for publication
40. H.K. Janssen: Z.Phys. 270, 67 (1974)
41. J. Oppenheim, K.E. Shuler, G.H. Weiss: Physica 88A, 191 (1977)
42. B.J. West, K. Lindenberg, V. Seshadri: J.Chem.Phys. 72, 1145, 1151 (1980)
43. L. Schimansky-Geier, W. Ebeling: Ann.Physik 40, 10 (1983)
44. R. Feistel: Nonlinear Chemical Reactions in Diluted Solutions, in this book
45. E.W. Montroll: in: Statistical Mechanics (Eds. A. Rice, K.F. Freed, J.C. Light) (University of Chicago Press, Chicago 1972)
46. G. Venzl: Phys.Rev. 31A, 3431 (1985)
47. M. Eigen: Naturwiss. 58, 465 (1971)
48. W. Ebeling, R. Feistel: Physik der Selbstorganisation und Evolution (Akademie-Verlag, Berlin 1982)
49. W. Ebeling, A. Engel, B. Esser, R. Feistel: J.Stat.Phys. 37, 369 (1984)

Dynamic Chaos in Ensembles of Structures and Spatial Development of Turbulence in Unbounded Systems

A.V. Gaponov-Grekhov and M.I. Rabinovich

Institute of Applied Physics, Academy of Sciences of the USSR, SU-Gorki, USSR

1. Development of Chaos in Ensembles of Dynamic Structures

1.1. Dynamic Structures

Development of spatial instabilities in nonequilibrium dissipative media is often a cause of regular formations which appear as ensembles of identical (or similar) elementary cells or dynamic structures. Rolls observed in thermoconvection in a plane-horizontal layer, Taylor vortices in the Couette flow between rotating cylinders, large vortices in underwater jets, wakes and other shear flows are examples of such dynamic structures. The onset of developed turbulence in a nonequilibrium medium is often preceded by several stages of gradual complication and transformation of the structures. For example, structures with individual or collective degrees of freedom excited against the background of a regular ensemble. First, as a rule, the structures begin to oscillate: for example, bending oscillations of convective rolls or azimuthal waves in Taylor vortices or in toroidal vortices in underwater jets appear. Then, with a further increase in supercriticality, the coupling between the neighbouring elements weakens and the resulting nonequilibrium medium can be considered (in a certain range of parameters) as a discrete ensemble of interacting oscillating structures.

1.2. Example. Chaotic Self-Modulation of Two-Dimensional Structures on the Surface of a Fluid at Parametric Excitation

Let us consider transition to turbulence via chaotic modulation of two-dimensional structures on the surface of a viscous fluid in a periodically oscillating homogeneous field of gravity /1/. Transition to chaos in our experiments was observed on a horizontal surface with about 10^4 cells-oscillators. Irrespective of the shape of the boundaries, turbulence occurred against the background of two-dimensional structures due to their self-modulation that was random in space and time. We used a layer of silicon oil 0.5 cm thick on a plane surface of a vibrator. The surface of the fluid was photographed in reflected light with subsequent space Fourier analysis of the image contrast using an optical spectrum analyser.

As the oscillation amplitude of the cells increased the following transitions were observed. A regular grating with square cells appeared at the first critical value of the vibration amplitude R_{cr_1}. The grating was formed by two pairs of parametrically excited travelling waves (photo a) in Fig. 1). These pairs of waves propagated orthogonally, irrespective of the shape of the cell boundaries, thus forming a tetrahedron.

A further increase in R resulted in the appearance of modulation waves with a space period of the order of the cell size. The direction of their propagation coincided with that of the parametrically excited waves forming the tetrahedron. With a still further increase

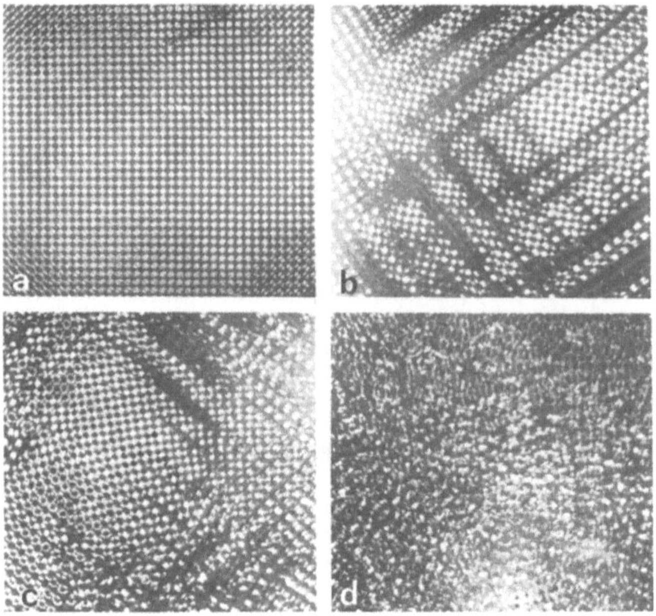

Fig. 1 Transition of chaos in an ensemble of structures on the surface of a parametrically excited fluid: a) $r_2 > r > r_1$ - square cells formed by two pairs of counterpropagating waves appear on the surface; b) $r_3 > r > r_2$ - modulation waves appear against the background of the ensemble of square cells; c) $r \gtrsim r_3$ - dislocations appear; d) $r > r_3$ - chaos onsets in the ensemble of cells /1/

in R, the velocity of the modulation waves increased and their space period Λ decreased. Transition to chaos occurred at $R = R_{cr_3}$, when the regime of shortest-wave modulation lost its stability. Straight after the critical point modulation acquired quite a specific character. Large blocks with regularly located cells (laminar sections) were observed, along with small sections of chaotically located cells-dislocations (turbulence). With an increase in supercriticality, the laminar sections reduced, while the turbulent onces enlarged. This is illustrated in Fig. 1. The photo b) corresponds to preturbulence regime ($r = (R-R_{cr_3})/R_{cr_3} = -0.2$), the photo c) corresponds to the critical point ($r = 0.05$) and, finally, the photo d) corresponds to developed chaos ($r = 1.4$).

1.3. Model

Phenomenological description of the observed transition to turbulence can be given using a discrete (in space) dynamic model. The elementary cells in such a model should be nonlinear parametrically excited coupled oscillators. Note, that numerical study of a similar model, a one-dimensional sequence of coupled self-excited oscillators, showed a transition analogous to the observed one. As in the experiment, stochasticity was preceded by a regular space-time self-modulation which transform to chaotic modulation via intermittency at $r = 0$.

Let us now consider the dynamics of an ensemble of structures. Assume that only one degree of freedom is excited in each structure

$$\frac{da_j}{dt} = a_j(1-\delta|a_j|^2) \ . \qquad (1)$$

Here $a_j(t)$ is a complex variable that characterizes the state of the structure and $\delta = 1+i\beta$ is a complex parameter which defines the amplitude and pulsation frequency of the structure in the stationary regime. A two-dimensional ensemble of such interacting structures is described by a differential equation

$$\frac{da_{jk}}{dt} = a_{jk}-(1+i\beta)|a_{jk}|^2 a_{jk} + \varkappa \hat{L}\, a_{jk} \ , \qquad (2)$$

where $\hat{L}$ is a linear differential operator. For a one-dimensional case we have

$$\frac{da_j}{dt} = a_j-(1+i\beta)|a_j|^2 a_j+e(1-ic)(a_{j+1}-a_{j-1}-2a_j) \ , \qquad (3)$$

where β, c, and e are larger than zero. We have restricted our analysis to ensembles with periodic boundary conditions: $a_j(t) = a_{j+N}(t)$, where N is the number of structures in the ensemble. We have shown /2/ that all collective motions in the form of stationary travelling waves are unstable in a certain parameter range, and quasi-periodic or chaotic motions occur in the ensemble. Together with I.S. ARANSON we have given the upper estimate for the entropy of the Kolmogorov-Sinai chaotic set and determined the fractal dimension of a strange attractor. We have found the dependence of the entropy and dimension on the coupling between the structures in the ensemble with an arbitrary (any finite) number of elements. The dependence of the strange attractor dimension on the number of unstable stationary waves available among the solutions has also been determined. The theory has been compared with the numerical experiment carried out earlier /3/.

1.4. Development of Chaos

The degree of evolution of multidimensional chaos can be determined by the following characteristics of stochastic sets:
by the normalized Kolmogorov-Sinai entropy

$$H_\lambda = \sum_{j=1}^{m} \lambda_j/\lambda_1; \qquad (4)$$

here λ_j are the Ljapunov indices in descending order $\lambda_1 > \lambda_2 > \ldots > \lambda_m > 0 > \ldots > \lambda_{2N}$), and by the attractor dimension

$$D_\lambda = M + d \qquad (5)$$

(here M can be found from the condition $\sum_{j=1}^{M}\lambda_j > 0, \sum_{j=1}^{M+1}\lambda_j < 0; d = \sum_{j=1}^{M}\lambda_j/|\lambda_{M+1}|$).

The entropy and dimension can be estimated by the mean eigenvalues $\sigma_j(t)$ of the auxiliary matrix $\bar{B}(t) = [B(t)-B^+(t)]/2$. Here B(t) is the matrix of the initial set (2) or (3) in a real form linearized in the vicinity of a typical solution that belongs to a stochastic set, $B^+(t)$ is the emitter-conjugate matrix. The sum of the first characteristic indices $h_l = \sum_{i=1}^{l}\lambda_i$ needed for the calculation of the entropy and dimension of the stochastic set is related to $\sigma_j(t)$ by

$$h_1 \leqslant \lim_{t\to\infty} \frac{1}{t} \int_0^t \sum_{j=1}^{l} \sigma_j(\tau)d\tau , \qquad (6)$$

where $\sigma_1(t) \geqslant \sigma_2(t) \geqslant \ldots \geqslant \sigma_e(t)$.

In view of translation symmetry of the sequence we restrict our consideration to space-homogeneous stochastic regimes $\langle |a_j^o|^2 \rangle = |a^o|^2$, $\langle a_j^{o\,2} \rangle_t = a^{o2}$. Besides, we assume that the mean intensity pulsations are weak, i.e. chaos is basically phase chaos: $|a_j^o|^2 - |a^o|^2 = z_s$ $(\langle z^2 \rangle)^{1/2} \ll |a^o|^2$. Numerical experiments confirm this assumption /3/.

Thus, we can readily find all eigenvalues of the matrix $\hat{B}(t)$

$$\sigma_{n+1} = 1-4e \sin^2 \frac{\theta_n}{2} + [(1+\beta^2)^{1/2}-2]\, |a^o|^2 ,$$

$$\sigma_{N+n+1} = 1-4e \sin^2 \frac{\theta_n}{2} - [(1+\beta^2)^{1/2}+2]\, |a^o|^2 , \qquad (7)$$

where $\theta_n = \pm 2\pi n/N$ $(n=0,1,\ldots,N/2)$. For strong coupling $(e > 1/2)$, the upper estimate for D_λ can be obtained directly from the equation

$$H_{D_\lambda}(e) = D_\lambda\,(1-2e) + 2e\,\frac{\sin(D_\lambda \frac{\pi}{N})}{\sin\left(\frac{\pi}{N}\right)} = 0 . \qquad (8)$$

For weak coupling $(e < 1/2)$, taking into account that $(|a^o|^2 \approx 1)$ we have

$$D_\lambda \leqslant N+n = N(1+(1-2e)/(4|a^o|^2-1)) \approx N(1+(1-2e)/3) . \qquad (9)$$

Figure 2 gives a comparison of the theoretical results and the numerical experiment.

1.5. Dimension Versus the Number of Unstable Waves

In order to elucidate the physical nature of growing dimension for stochastic motions with an increase in autonomy of oscillating structures, i.e. with a decrease in e, we analyze the feature of regular collective excitations in the form of stationary travelling waves

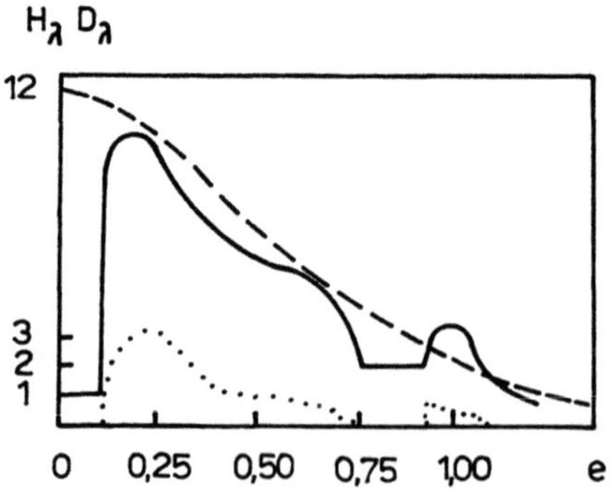

Fig. 2 Comparison of the estimate D_λ with numerical experiment for a one-dimensional sequence with $\hat{N}=9$ ($\beta = c = 1.71$): the solid curve is given for the numerical experiment /3/, the broken curve, for the analytical estimate /2/

$$a_j(t) = A_n \exp[i(\omega_{n}t + j\,\theta_n)] \; . \qquad (10)$$

The dependence of the stationary wave intensity on the propagation constant θ_n can be written in the form $|A_n|^2 = 1-4e \sin(\theta_n/2)$ and the law of dispersion for these waves, in the form $\omega_n = -\beta+4e(\beta+c)\cdot \sin^2(\theta_n/2)$. Here $\theta_n = \pm 2\pi n/N$ and $n=0,1,\ldots,N/2$. The stability of the space-homogeneous regime is defined by

$$\lambda_{\ell_o} = 1 - 4e \sin^2 \frac{\theta_e}{2} \pm [1-(4ce \sin^2 \frac{\theta_e}{2})^2+8\beta ce \sin^2 \frac{\theta_e}{2}]. \qquad (11)$$

Hence, for

$$e > e_o = (\beta c - 1)/ [2(1+c^2) \sin^2 \frac{\pi}{N}], \qquad (12)$$

there are no negative indices ($\lambda_{l_o} \lesssim 0$) and the regime of space-homogeneous oscillations in the ensemble of N cells is stable.

For $\beta c > 1$, a decrease in e from e_o results in the appearance of new stationary waves (10) with n=1,2,...,N/2 from a trivial equilibrium state a=0 at $e=e_n$.

In the phase space of set (3) this corresponds to gradual onset of pairs of limit cycles (for forward and counterpropagating waves) from the equilibrium state at e_n. All periodic motions thus born are unstable.

Let us assume now that all newly born motions belong to a strange attractor. Then we can readily interpret the results given above for monotonic increase in the chaos dimension with increasing autonomy of the structures. As e decreases, trajectories appear in the attractor. The number of unstable directions for these trajectories is of the order of the number of the new cycle (periodic motion with $\theta = \theta_e$).

Thus, the number of unstable periodic solutions of type (10) existing at a given e is the lower estimate of the attractor dimension.

2. Spatial Development of Chaos in Sequencies of Nonreciprocally Coupled Structures

2.1. Formulation of the Problem. Models

The nature and mechanisms of spatial development of turbulence are complicated problems in stochastic behaviour of nonequilibrium dissipative media. These problems are essential for the understanding of the origin of turbulence in shear flows (jets, wakes, shear layers etc.). Why and how does the perturbation that is harmonic at the beginning of the system get more and more complicated and, finally, become completely chaotic in the course of spatial evolution? Is there any connection between spatial development of chaos in a semi-infinite dissipative system and strange attractors? We believe that these questions may be answered if we thoroughly investigate the development of perturbations and "spatial bifurcations" of appearing chaos in semi-infinite sequences of structures. A convective character of interaction between the structures (oscillating vortices) in such sequences is simulated by nonreciprocal coupling of elements in the sequence. A general model can be written in the form

$$\frac{d\vec{u}_j}{dt} = F(\vec{u}_j)+\gamma(\vec{u}_j-\vec{u}_{j-1})+ \varkappa(\vec{u}_{j+1}-2\vec{u}_j+\vec{u}_{j-1}) \; , \qquad (13)$$

where $F(\vec{u}_j)$ characterizes the dynamics of an elementary cell (structure), γ is responsible for unidirectional effect of the (j-1)-th cell on the j-th one (simulation of a convective character of the bifurcation evolution along the sequence); $\varkappa$ is a weak feedback - between the structures "upstream" (diffusion). The early results on the study of such systems were obtained in /4/. We also accomplished numerical experiments on unidirectional sequences of successively coupled oscillators for $\varkappa=0$. Such sequences simulate a semi-infinite nonreciprocal medium. Harmonic oscillations in systems of this type develop "downstream" into quasiperiodic and then chaotic oscillations with a finite /4/ or linearly growing /5/ dimension. In particular, in semi-bounded nonreciprocal media whose field is described by a discrete model

$$\frac{da_j}{dt} = a_j(1-\delta|a_j|^2) - \gamma(a_j-a_{j-1}) , \quad j = 1,2,... \tag{14}$$

turbulence developing along j may onset at $a_1(t)-a_0(t)\equiv 0$. Set (14) was studied numerically at $\gamma = \alpha'(1-i\alpha'')$, $\delta = 1-i\beta$, where $\alpha', \alpha'', \beta > 0$. The power spectra of amplitude and phase $a_j(t)$ were analyzed and the dimension of the realization of $a_j(t)$ "downstream" was calculated as a function of j (Fig. 3).

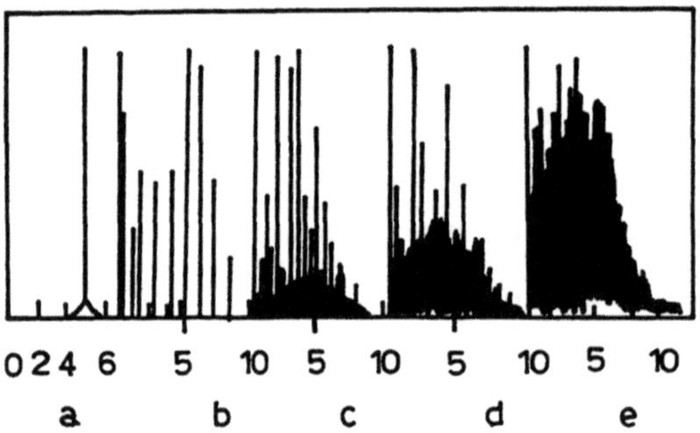

Fig. 3 Power spectra of $a_j(t)$ illustrating spatial development of turbulence along the sequence: a) j=2, b) j=9, c) j=10, d) j=12, e) j=20 and 50; $\alpha'=0.5$; $\alpha''=1.71$; $\beta=5.0$ /4/

It may seem that the dimension of the strange attractor should increase "downstream", since owing to nonreciprocal coupling the number of degrees of freedom that take part in the motion increase with the growth of j. This means that "infinitely dimensional" turbulence could be expected at $j\rightarrow\infty$. However, this is not the case even in a homogeneous sequence (γ, δ =const.). Spatial development of turbulence occurs in the sequence. At small j we deal with a quasi-monochromatic regime, then with an increase in j it turns into a regime of pulsations with a large number of harmonics and, finally, with a further motion "downstream" it becomes weakly turbulent. Turbulence develops, its power increases, though it does not change any longer at sufficiently large; i.e. a regime of stationary space-homogeneous turbulence is established.

2.2. Physical Experiments on Spatial Development of Chaos in Unidirectional Sequences of Coupled Oscillators

A sequence of oscillators with inertial nonlinearity /6/ was studied. Each oscillator in autonomous regime allowed transition to stochasticity via period doubling demonstrating Feigenbaum's universality. It was found that transition to chaos along the sequence was accompanied by several period doublings /7/. For a small degree of nonequilibrium K, synchronous generation of oscillations with a period $2^m T_0$ was established along the sequence after several multiplications on the first oscillators. Further increase in K resulted in a chaotic regime along the sequence, with a spectrum width and, probably, dimension being constant "downstream". This phenomenon of reemission of a chaotic signal without variation of the spectrum can be named chaotic synchronization. The basic effects in the sequence observed experimentally by V.S. ANISHCHENKO and D.E. POSTNOV are shown in Fig. 4. Here the onset of different regimes along the sequence is given versus K.

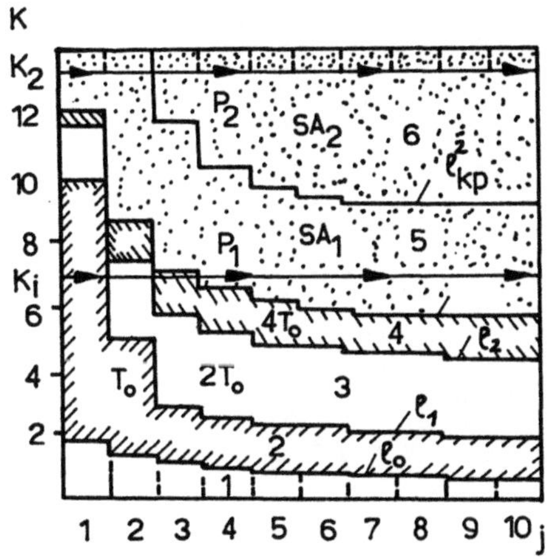

Fig. 4 Bifurcation diagram for various regimes of oscillations in the plane of K and j parameters. Physical experiment /7/

In region 1 the system is not excited and is essentially a sequence of coupled regenerative contours. Transition across the boundary l_0 corresponds to soft excitation of self-oscillations and the system has regular oscillations with a period T_0 in region 2. The increase of that corresponds to transition across l_1 and l_2 results in in the onset of double-period oscillations: $T_0 \xrightarrow{1} 2T_0$, $2T_0 \xrightarrow{2} 4T_0$. Further increase of the degree of nonequilibrium, i.e. transition across l^1_{cr}, results in the onset of chaotic oscillations. The image of these oscillations is represented by a strip strange attractor with a small fractal dimension. Their energy spectrum has clearly defined peaks at the fundamental frequency $f_0 = 1/T_0$, half-frequency $1/2\ f_0$ and at their harmonics. The peaks disappear as a result of transition across l^2_{cr}, which corresponds to the appearance of another attractor.

The following regime were established along the sequence at fixed K: a) spatial synchronization of oscillations at the fundamental frequency $(1<K<2)$; b) spatial synchronizations of oscillations with a period $2T_0$ $(3<K<4.5;\ j\geq 3)$; c) spatial synchronization with a period $4T_0$ $(4.5<K<5.5;\ j\geq 5)$; d) homogeneous spatial chaos, or chaotic synchronization in space $(K\approx 7,\ j\gtrsim 3)$; the energy spectrum of stochastic oscillations did not depend on j. The evolution of oscillations along the sequence at K=7.0 is seen in Fig. 5. Here the oscillation energy spectra in elements 1-4,6 and the phase portrait of the strange attractor for j=6 are shown.

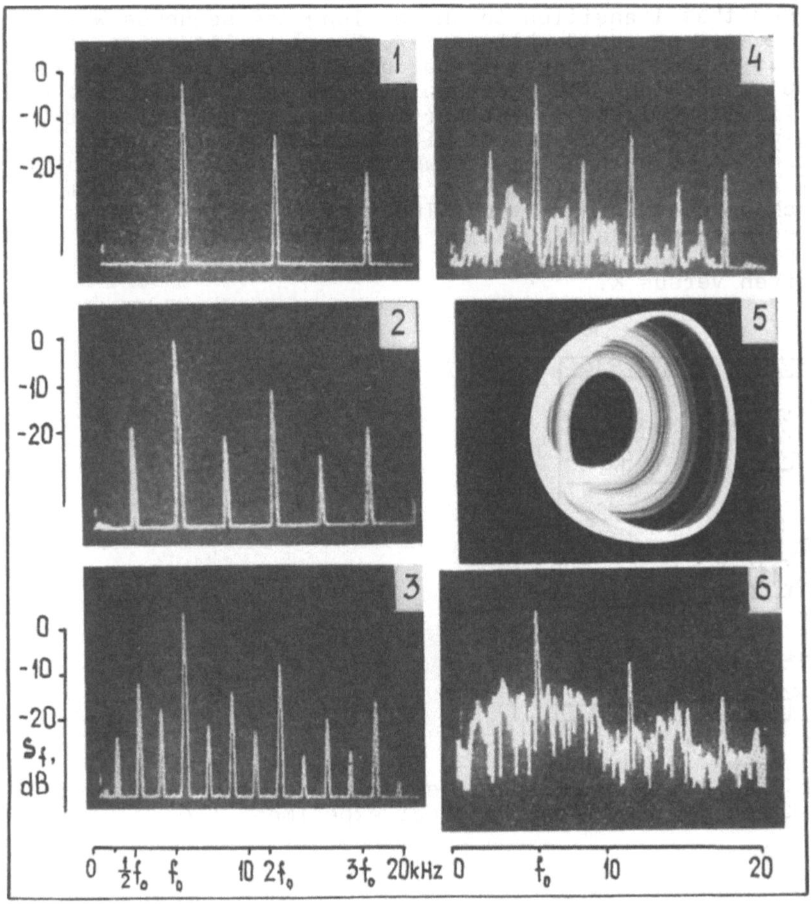

Fig. 5 Evolution of power spectra in transition to chaos via spatial bifurcation sequence: 1) T_0 cycle, j=1; 2) $2T_0$ cycle, j=2; 3) $4T_0$ cycle, j=3; 4) strange attractor SA_1, j=6; 5) SA_1, projection of the phase trajectory on the X_6Y_6 plane; 6) developed stochasticity of SA_1, j=10. (Photographs taken by V.S. ANISHCHENKO and D.E. POSTNOV)

2.3. Model Theory

As supercriticality increases, each oscillator in the sequence demonstrates transition to chaos via period doubling. Systems with such a transition to chaos are described by a parabolic one-dimensional map /8/. Hence, to understand and describe the phenomena observed in the sequence, we consider a model medium in the form of maps coupled unidirectionally /7/

$$x_j(n+1) = 2Kx_j(n) + 2x_j^2(n)+qx_{j-1}(n+1) \; ; \; x_0(n) = 0 \; ; \; j=1,2,\ldots \qquad (15)$$

where n is a discrete time, $K>0$ characterizes the degree of nonequilibrium of the system, and $q\in(0,1)$ is the coupling. It is easy to explain the character of transitions in the sequence on the way to chaos.

Spatial transitions observed at not too large K (see Fig. 4) can be analyzed by the evolution of fixed points of individual maps along the sequence. Assume that regular oscillations with a period T_0 are established in the first oscillator. This regime corresponds to a stable fixed point of an individual map. This is possible at $1/2 \leqslant K \leqslant 3/2$. Assuming $x_j(n+1) = x_j(n)$, from (15) we find a one-dimensional map that describes an inhomogeneous regime of fixed points.

$$x_j^o = 2Kx_j^o + 2x_j^{o2} + qx_{j-1}^o . \tag{16}$$

This map is single-valued for $x_j^o, x_{j-1}^o < 0$. It is seen that at $j \to \infty$ the regime is approaching a spatially homogeneous regime, which corresponds to the tendency to a fixed point $x_o = \frac{1-q}{2} - K$. Linearizing (15) in the vicinity of regime (16), we find that this single-frequency regime is stable if the condition $|2K+4x_j^o| \leqslant 1$ is fulfilled for all j, i.e. a simple regime of spatial synchronization is realized when the coupling is stronger than the critical value $q > q_{cr} = 3/2-K$. If $q < q_{cr}$, then the fixed point loses its stability for a certain $j=j^*$. In view of the squareness of the individual maps, the T_0 period regime changes for the 2^mT_0 period regime. Here m=1,2,... depends on the degree of nonequilibrium and on the value of coupling. If the coupling is weak, period doubling will be the first transition across the boundary. We then consider the j-th oscillator same as the first one. We easily find the conditions for synchronization at the $2T_0$ frequency and so on.

It can be shown that when the regime of spatial synchronization of regular oscillations is not realized in the sequence and chaos is established at $j \to \infty$, the number of doublings on the way to oscillator /8/. This can be explained by the fact that each period doubling bifurcation of oscillations along the sequence is accompanied by effective two-fold increase of coupling between the neighbouring oscillators.

3. Conclusions

The estimates obtained for the entropy and dimension of chaos that is realized in nonequilibrium media under study enable us to predict the dependence of some characteristics of turbulent regimes on parameters and to understand physical mechanisms of developing chaos. A close relation between the dimension of chaos and stationary waves means that a change in parameter the instant a new stationary wave appears results in a new unstable direction of trajectories belonging to the strange attractor. Thus the parameters of unstable stationary waves, such as their amplitudes and phases, can be considered as new usual variables whose number roughly defines the dimension of chaos in an ensemble of structures. This estimate can also be used for a continuum system. Note,that the dimension of chaos in this case can be expressed in terms of the system length, and the connection between the dimension and the effective Reynolds number can be readily traced.

In a nonreciprocal nonequilibrium sequence, a "medium" that allows the propagation of perturbations only in one direction, turbulent regime may onset as a result of a sequence of spatial bifurcations, same as occurs in a point oscillator when its parameter is changed. The fact that the dimension and the statistical features of the chaotic regime do not depend on the coordinate at large j (effect of chaotic synchronization) means that turbulence is developing into a stationary regime characterized by partial coherence. Turbulence is finite-dimension even at $j \to \infty$. We believe that the mechanisms according to which turbulence is stabilized in homogeneous nonequilibrium media, as well as the estimation of the variations in the strange attractor dimension, are of undoubted interest. We can add to it that according to numerical experiments low diffusion does not indeed qualitatively influence the effects observed in convective (unidirectional) sequences.

References

1. A.B. Ezersky, P.I. Korotin, M.I. Rabinovich: Chaotic self-modulation of two-dimensional structures on the fluid surface at parametric excitation, Pisma v Zh.Eksp.Teor.Fiz. 41, 129 (1985)
2. I.S. Aranson, A.V. Gaponov-Grekhov, M.I. Rabinovich: Development of chaos in ensembles of dynamic structures, Zh.Eksp.Teor. Fiz. 89, 92 (1985)
3. A.V. Gaponov-Grekhov, M.I.Rabinovich, I.M. Starobinets: Appearance of multidimensional chaos in active gratings, Dokl. Akad.Nauk SSSR 279, 596 (1984)
4. A.V. Gaponov-Grekhov, M.I. Rabinovich, I.M. Starobinets: Dynamic model of spatial development of turbulence, JETP Lett. 39, 688 (1984), Pisma v Zh.Eksp.Teor.Fiz. 39, 561 (1984)
5. V.S. Afraimovich, M.I. Rabinovich, B.I. Sbitnev: Attractor dimension in a sequence of coupled oscillators, Pisma v Zh.Teor. Fiz. 6, 338 (1984)
6. V.S. Anishchenko, V.V. Astakhov: Experimental study of the mechanism of appearance and the structure of a strange attractor in a generation with inertial nonlinearity, Radiotekhnika i Elektronika 28, 1109 (1983)
7. V.S. Anishchenko, I.S. Aranson, D.E. Postnov, M.I. Rabinovich: Spatial synchronization and bifurcations of developing chaos in a sequence of coupled oscillators, Dokl. Akad. Nauk USSR 286, 596 (1986)
8. A.J. Lichtenberg, M.A. Lieberman: Regular and Stochastic Motion (Springer, Berlin Heidelberg New York 1983)

Part II

Stochastic and Statistic Theory

Nonlinear Chemical Reactions in Diluted Solutions

R. Feistel

Humboldt-Universität Berlin, Sektion Physik, Invalidenstraße 42, DDR-1040 Berlin, GDR

In the theory of selforganization we have to describe some fundamental and nontrivial phenomena /1,2,3/:
(i) Nonlinearity of the kinetic equations, caused by far-from-equilibrium conditions and the existence of feedback loops in the system, giving rise to kinetic phase transitions, multistability, selfsustained oscillations or even chaotic behaviour;

(ii) Spatial transport, e.g. via diffusion or convection, leading to spatial patterns and long-range correlations, to moving phase fronts and nucleation phenomena;

(iii) Noise, e.g. thermal noise or shot noise of chemical reactions, requiring probabilistic descriptions and bringing noise-induced phenomena into the picture;

(iv) Interacting particles like ions or hard spheres, being responsible for electrical dissipative structures or nonideality effects.

As a rule, models of selforganization deal usually only with one or two of these properties at the same time. A possible tool to unify the different aspects mentioned is the functional master equation (FME)

$$\frac{\partial}{\partial t} P(n(r),t) = \underline{W}\, P(n(r),t) \quad , \tag{1}$$

describing the evolution of a functional probability density P and possible density profiles n(r) undergoing changes by transport and chemical processes given by the operator $\underline{W}$. Here we shall consider only systems with one chemical component in a diluted solution, being in thermal equilibrium with the bath liquid.

To construct the evolution operator $\underline{W}$ we summarize first some necessary requirements:
(i) Nonideality should enter into the description by arbitrary pair interaction potentials Ψ_{ij} between the particles.

(ii) Chemical reactions should appear as jumps in a discrete particle number space, generalizing the usual master equation formalism.

(iii) Diffusion should be described in a continuous way following the scheme of Fokker-Planck equations.

(iv) Hydrodynamics should be represented by a solvent velocity field $u_0(r)$ that might be subject to a Navier-Stokes equation.

(v) Reactive collisions should be governed by specific configuration-dependent reaction probabilities.

(vi) Classical statistical transport theory should be the limiting case of the FME in absence of chemical reactions.

This last condition may serve as a starting point for deriving the FME formalism.

Equilibrium and transport properties of diluted isothermal solutions can be calculated from the statistical diffusion equation of STRELCOVA /4/ and EBELING /5/:

$$\frac{\partial}{\partial t} D_N + \sum_i \frac{\partial}{\partial r_i} \{ u_i D_N \} = 0$$

$$u_i = u_o(r_i) - \frac{1}{\varrho_i} \frac{\partial}{\partial r_i} \{kT \ln D_N + U(r_i) + \sum_j \Psi_{ij} (r_i r_j)\}$$

$$- \frac{1}{\varrho_i} \sum_k \chi_{ik} \frac{\partial}{\partial r_k} \{kT \ln D_N + U(r_k) + \sum_j \Psi_{kj}(r_k r_j)\} . \qquad (2)$$

Here $D_N(r_1,r_2,...,r_N,t)$ is the statistical N-particle distribution function for the probability density to find particle 1 at r_1... r_1+dr_1 etc., u_i is the velocity of the i-th particle, $u_o(r)$ the velocity of the solvent at r, ϱ_i the friction coefficient. On particle i there are acting the diffusion force, an external force with the potential U(r) and the other particles j with potential Ψ_{ij}. χ_{ik} is the Oseen coupling tensor for the influence of a moving particle k on particle i via hýdrodynamic interaction.

In (2) averages have been carried out over the momentum space as well as over the fast thermal motion of the solved particles and the solvent supposing thermal equilibrium. The solvent appears as a continuum characterized by friction coefficients ϱ_i, its temperature T, its viscosity (in χ_{ik}), and its dielectric constant (in Ψ_{ij}) if Coulomb forces are present.

This transport equation is valid for a system with a fixed number N of solved particles. If we include also chemical reactions, this number will no longer be fixed and we introduce P(N,t) for the probability having just N particles in the system. To have, in addition, the information about their spatial distribution, we write $P(N;r_1...r_N,t) = P(N,t) \cdot D_N(r_1,r_2,...,r_N,t)$. It will change in time due to transport and chemical processes

$$\frac{\partial}{\partial t} P(N;r_1...r_N,t) = (\frac{\partial}{\partial t})_{tr} P + (\frac{\partial}{\partial t})_{ch} P$$

$$= P(N,t) \frac{\partial}{\partial t} D_N + \underline{W}_{ch} P(N;r_1...r_N,t) . \qquad (3)$$

The transport contribution is given by the diffusion equation (2); the chemical part will be constructed in a way like the usual chemical master equation for discrete particle numbers. What we get is a mixed continuous (in space) and discrete (in particle numbers) description that may serve as a starting point to study spatial correlations of chemically reacting and physically interacting (nonideal) particles influenced by a given hydrodynamic motion of the solvent. The description is consistent with classical statistical theories, for the transport equation reduces to the well-known BBGKY-hierarchy under equilibrium conditions. Equation (3) may be rewritten using instead of $P(N;r_1...r_N,t)$ the functional probability density $P(n(r),t)$ with $n(r) = \sum_i \delta(r-r_i)$. Both are equivalent but the functional representation is less worth for concrete calculations, although it reflects the relationship to the ordinary master equation more clearly: instead of adding (or substracting) 1 to the particle number N we have to add (or remove) $\delta(r-r_i)$ to the distribution $n(r)$.

Following the line of master equations for chemical reactions, we shall assume that the reactions can be taken as a Markov process.

Let the birth rate of a particle at r be given by $W_+(r|N;r_1...r_N)$, depending on the positions r_i of all present N particles. Similarly $\underline{W}\,(r_i|N;r_1...r_N)$ is the rate that just the i-th particle of N existing disappears by a death process. Then the probability balance is

$$\begin{aligned}\underline{W}_{ch}P \equiv \frac{1}{NV} &\sum_i W_+(r_i|N-1;r_1...r_{i-1},r_{i+1}...r_N)P(N-1;r_1...r_{i-1},r_{i+1}...r_N,t)\\ &+ \int dr_{N+1}W_-(r_{N+1}|N+1;r_1...r_{N+1})P(N+1;r_1...r_{N+1},t)\\ &- \frac{1}{V}\int dr_{N+1}W_+(r_{N+1}|N;r_1...r_N)P(N;r_1...r_N,t)\\ &- \frac{1}{N}\sum_{i=1}^{N} W_-(r_i|N;r_1...r_N)P(N;r_1...r_N,t)\ . \end{aligned} \tag{4}$$

We have to relate now the $W_\pm$ to the macroscopic chemical reaction velocities.

A single reaction event concerns only a few (normally only one or two) of all present particles, therefore we may decompose $W_\pm$ into different-s-particle reactions

$$W_\pm = \sum_s W_s^\pm\ ,$$

$$W_s^+(r|N;r_1...r_N) = V\,k_s^+\sum{}^*\, C_{s+1}(r_1'...r_s'r)\ ,$$

$$W_s^-(r_i|N;r_1...r_N) = N\,k_s^-\sum{}^*\, C_s(r_1'...r_s'r_i)\ ; \tag{5}$$

k_s are the chemical rate constants, and $\sum^*$ is to read as the sum over all $N!/(N-s)!$ different choices of s particles (at r_i')of N (at r_i). In the case of W_s^-, choose only (s-1) of N and avoid particle i.

The reaction rates k_s give the chance that somewhere in the volume V a reaction takes place at the given time t...t+dt with any configuration of the reacting particles. Consequently, when a reaction happens, then with a certain chance in a certain s-particle configuration. This conditional probability density is denoted by C_s/V. It follows that

$$C_0 \equiv V\ ;\quad \int dr_1...dr_sC_s(r_1...r_s) = V \tag{6}$$

or less generally

$$c_1 \equiv 1\ ;\quad \int dr_1...dr_{s-1}C_s(r_1...r_s) = 1 \tag{7}$$

under homogeneous reaction condition, where the reaction depends only on the relative positions of the particles but not on the absolut localization in the reactor.

The time intervals dt we consider here are long compared with the molecular collision frequency. So C_s gives the chance of a reactive collision within this time interval.

In the following we shall consider the functions C_s as empirically given. They reflect somehow the detailed microscopic mechanism of the reaction as e.g. the fast thermal motion with elastic and reactive collisions, fast intermediate reactions, building of complexes etc. Nevertheless the functions C_s can be determined physically in special cases, e.g. the ionic association reaction in an electrolyte.

Let us briefly discuss some simple situations.

If all reactions take place independently of the particle configuration (i.e. C_s = const) then (1) reduces to an ordinary chemical master equation.

If, to the other extreme, we have C_s strictly δ-like localized (i.e. particles react only when they are in identical positions), then nonlinear reactions produce divergent behaviour.

Another possibility is taking C_s as constant when all vectors $r_1 \ldots r_s$ point into the same space cell, and zero otherwise. Then (4) becomes equivalent to the chemical part of the multivariate master equation /6/.

To make the equations more transparent, we shall take as an example for the chemical rate equations the quadratic SCHLÖGL model reaction

$$0 \underset{B}{\overset{A}{\rightleftharpoons}} X \underset{C}{\overset{E}{\rightleftharpoons}} 2X \tag{8}$$

corresponding to a rate equation

$$\dot{X} = f(X) = A-BX+EX-CX^2 . \tag{9}$$

To construct the space-dependent birth and death rates (5) we start from the homogeneous transition rates /7/

$$W_+(N) = AV + EN ,$$

$$W_-(N) = BN + CN(N-1)/V . \tag{10}$$

Supposing homogeneous conditions, we have $C_1(r) = 1$ for unimolecular reaction (as well as import or export of X). Then we use $C_2(r_i, r_j)$ for the description of binary reactive collisions

$$W_+(r|N,r_1\ldots r_N) = AV + EV \sum_{j=1}^{N} C_2(rr_j) = W_0^+ + W_1^+ ,$$

$$W_-(r_i|N;r_1\ldots r_N) = BN + CN \sum_{\substack{j=1 \\ j\neq i}} C_2(r_i r_j) = W_1^- + W_2^- . \tag{11}$$

Inserting these rates into the chemical part (4) as well as the transport equation (2) we get the complete master equation explicitly. This is of course a rather long expression, which can be written more compactly using the generating functional

$$G(s(r),t) = \sum_N \int dr_1 \ldots dr_N \; s(r_1)\ldots s(r_N) P(N;r_1\ldots r_N,t) . \tag{12}$$

Dropping for simplicity all hydrodynamic contributions, one gets

$$\frac{\partial}{\partial t} G(s(r),t) = \int dr_1 \left[s(r_1)-1\right]$$

$$* \Big\{ D\frac{\partial^2}{\partial r_1^2}\frac{\delta G}{\delta s(r_1)} + \frac{1}{\varrho}\frac{\partial}{\partial r_1}\frac{\partial U}{\partial r_1}\frac{\delta G}{\delta s(r_1)} + \frac{1}{\varrho}\frac{\partial}{\partial r_1}\int dr_2 s(r_2)\frac{\partial \Psi(r_1 r_2)}{\partial r_1}\frac{\delta^2 G}{\delta s(r_1)\delta s(r_2)}$$

$$+ AG - B\frac{\delta G}{\delta s(r_1)} + E\int dr_2 C_2(r_1 r_2)s(r_2)\frac{\delta G}{\delta s(r_2)} - C\int dr_2 C_2(r_1 r_2)s(r_2)\frac{\delta^2 G}{\delta s(r_1)\delta s(r_2)}\Big\} \tag{13}$$

with the diffusion coefficient $D = kT/\varrho$.

As is well known from statistical physics near equilibrium, it is not possible to solve the full N-particle problem (13) in the stationary case for nontrivial problems - the same applies here. To calculate thermodynamic and transport properties one usually splits the equation for D_N into a hierarchy of coupled equations for the reduced distribution functions F_s by integration over the remaining (n-s) coordinates. Following this route for the master equation, we get a network of such hierarchies of coupled equations for the functions $F_s^N \equiv V^s \int dr_{s+1} \dots dr_N P(N,r_1 \dots r_N,t)$:

$$\begin{array}{lccccccc}
 & N=0 & & N=1 & & N=2 & & \dots \\
s=0 & P(0) & — & P(1) & — & P(2) & — & \dots — P(N) \\
 & & & | & & | & & | \\
s=1 & & & F_1^1(r) & — & F_1^2(rr') — & & \dots — F_1^N \\
 & & & & & | & & | \\
s=2 & & & & & F_2^2(rr') — & & \dots — F_2^N \\
 & & & & & & & \vdots \\
 & & & & & & & F_s^N \\
 & & & & & & & \vdots
\end{array} \tag{14}$$

The equation for P(N) does not contain space coordinates explicitly, therefore the coupling with the rest of the network is only of algebraic nature and does not produce serious difficulties. To decouple the equations for F_s^N and F_{s+1}^N (vertically in our scheme) one makes use of common superposition approximations for the higher distribution functions. Finally to decouple F_s^N and F_s^{N+1} (horizontally) we shall assume that the N-dependence of the problem is reflected mainly via P(N) such that $F_s^N \sim F_s^{N+1}$ in those ranges of N where P(N) is significantly different from zero. This approximation is justified when N is not too small.

To get actually a row of the scheme (14) for a certain s one takes the s-th functional derivative of (13). Comparing the powers of s their coefficients give the equations for F_s for they have to vanish separately.

Beginning with s=0 we get a stationary master equation for P(N):

$$0 = AV\left[P(n-1)-P(N)\right] - B\left[NP(N)-(N+1)P(N+1)\right]$$

$$+ E\left[(N-1)P(N-1)-NP(N)\right] - C\,\frac{N}{V}\left[I(N)\cdot(N-1)P(N)-I(N+1)\cdot(N+1)P(N+1)\right] . \tag{15}$$

This equation differs from the homogeneous one only by the function

$$I(N) = \frac{1}{V}\int dr_1 dr_2 C_2(r_1 r_2) F_2^N(r_1 r_2) \ , \tag{16}$$

which expresses the influence of spatial correlations on the particle number distribution. This distribution is, applying the detailed balance condition to (15), given by the recurrence relation

$$P(N+1) = P(N)\ \frac{V}{N+1}\ \frac{AV+EN}{BV+CN\cdot I(N)}\ . \qquad (17)$$

For s=1 we get an equation for the one-particle distribution function $F_1(r)$. Neglecting the boundary influence on the system and assuming homogeneous reaction conditions ($C_1(r) = 1$; $C_2(r_1,r_2) = C_2(r_1-r_2)$) the spatial symmetry requires $F_1(r) = 1$. Considering the equation for F_1^N one can show that $F_1^N = 1$ is a solution, after some manipulations.

The equation for the pair distribution function $F_2^N(r)$, $r=r_1-r_2$, is already rather complicated. Setting $F_s^{N\pm1} \sim F_s^N = F_s$, dropping external and interaction forces, and applying unsymmetric superposition approximations

$$F_3(rr') \sim F_2(r)F_2(r');\ F_4(rr'r'') \sim F_2(r)F_2(r')F_2(r'')\ , \qquad (18)$$

we get a closed linear equation for F_2.

Now we introduce the correlation function $h(r) = F_2(r)-1$ and make use of the detailed balance (17) to end up with

$$\begin{aligned} &D\Delta h - \left[B + C\,\frac{N-2}{V}\,I + C\,C_2(r)\right] h(r) \\ &+ E\,\frac{N-2}{V}\,\frac{B+CnI}{A+En}\int dr'C_2(r-r')h(r') \\ &= (1-I)\,\frac{C}{V} + \left(\frac{1}{V} - C_2(r)\right)\frac{EB-AC+(I-1)E\ Cn}{A+En}\ , \end{aligned} \qquad (19)$$

where n abbreviates $n = (N-1)/V$.

In chemical equilibrium we have EB=AC, and the solution is trivial, h=0, I=1. In this special case we find no spatial correlation, and the particle number distribution is Poissonian

$$B(N+1)\ P(N+1) = A\ V\ P(N). \qquad (20)$$

The same is true when there is only one of both reactions, either A=B=0 or E=C=0. Finally, we get the trivial solution also in the perfectly mixed reactor, where $C_2(r) \sim 1/V$.

The diffusion equation (19) determines the correlation function h(r) which reflects the spatial correlation caused by local nonlinear chemical reaction. These correlations enter into the particle number distribution only via the function

$$I(N) = 1 + \int drC_2(r)h(r)\ . \qquad (21)$$

To find out the role diffusion may play, we can look at the possible range I(N) can vary when $C_2(r)$ or the diffusion constant are changed; therefore we multiply (19) by h(r) and integrate over the space coordinate.

After some manipulations we get from (19) finally the condition

$$(I-1)(EB-ACI) \geqslant 0\ , \qquad (22)$$

which limits I to the interval between 1 and EB/AC. As we know, I=1 is the case of vanishing spatial correlations ($D \rightarrow \infty$). The expectation that I=EB/AC is approached in the opposite situation, $D \to 0$, can be verified by looking in more detail at the derivation of (22).

What we found is that, for $D \to \infty$ all spatial correlation disappear (like in equilibrium), i.e. the correlation function has infinite range but zero amplitude, and I=1 gives a nonpoissonian particle number distribution. $D \to 0$, $I \to EB/AC$ produces a Poissonian particle number distribution (like in equilibrium) which is realized by strong spatial correlation over a very small range. Hence we have an interesting complementarity of the particle number distribution on one hand and the spatial distribution on the other hand. Varying the diffusion constant, we may bring one of them to its equilibrium shape, while the nonequilibrium character is reflected in the other distribution. Bringing both to equilibrium form requires linear chemical rate laws, or, equilibrium relations between the different rate constants.

One may believe that this complementarity is not restricted to the specific and simple model treated here, but reflects a general property of reaction-diffusion interaction.

In many situations reaction-diffusion systems are only weakly inhomogeneous, e.g. if the reactor is nearly perfectly stirred. Then one may be interested in only the first correction to the homogeneous case. This can be achieved in a long-range approximation for the reaction probability $C_2(r)$, giving rise to essential mathematical simplifications. We write

$$C_2(r) = \frac{1}{V} + \varepsilon(r) \; ; \quad \int \varepsilon(r)dr = 0 \tag{23}$$

for a small deviation $\varepsilon(r)$ from the homogeneous case. Then it follows that

$$h(r) = O(\varepsilon) \; , \; I = 1 + \int \varepsilon(r) \, h(r)dr = 1 + O(\varepsilon^2) \tag{24}$$

and we can linearize (19) with respect to ε :

$$D \, \Delta h - (B+nC) \, h = \varepsilon(r) \, \frac{AC-EB}{A+nE} \, . \tag{25}$$

This equation is easily integrated for any $\varepsilon(r)$:

$$h(r) = \frac{EB-AC}{(A+nE)(B+nC)} \, \frac{\varkappa^2}{4\pi} \int dr' \, \varepsilon(r') \, \frac{e^{-\varkappa|r-r'|}}{|r-r'|} \quad , \tag{26}$$

with $\varkappa^2 = (B+nC)/D$.

$\varkappa$ is the reciprocal diffusion length, i.e. the mean path length a particle is drifting between birth and death. With (24) we get for I(N)

$$I(N) = 1 + \frac{EB-AC}{(A+nE)(B+nC)} \, \frac{\varkappa}{4\pi} \int drdr' \, \varepsilon(r) \, \varepsilon(r') \, \frac{e^{-\varkappa|r-r'|}}{|r-r'|} . \tag{27}$$

Depending on N and D, (I-1) has always a definite sign

$$\begin{aligned} I(N) &> 1 \quad \text{if} \quad EB > AC \; , \\ I(N) &< 1 \quad \text{if} \quad EB < AC \; , \end{aligned} \tag{28}$$

which expresses the fact that, relative to the homogeneous case, weaker diffusion can on average only inhibit the nonlinear reactions, and never catalyze. I(N) shifts the mean value of P(N) in direction to the steady state of the linear reaction, as long as D is not too small. For very small D, I may become wrong expressing

that the linearized equation (25) fails for stronger spatial correlations. The limit of validity of (26) is indicated when I(N) violates the inequality (22).
As an example, we take $C_2(r) = \alpha^2 \exp(-\alpha r)/4\pi r$ with the spatial range $a=1/\alpha$.
We get the correlation function

$$h(r) = \frac{AC-EB}{(A+nE)(B+nC)} \left\{ \frac{1}{V} + \frac{\varkappa^2}{4\pi(1-\varkappa^2 a^2)} \left(\frac{e^{-\alpha r}}{r} - \frac{e^{-\varkappa r}}{r} \right) \right\} . \tag{29}$$

and the inhibition function I(N)

$$I(n) = 1 + \frac{AC-EB}{(A+nE)(B+nC)} \left\{ \frac{1}{V} - \frac{\alpha \varkappa^2}{8\pi(1+\varkappa a)^2} \right\} . \tag{30}$$

To get an analytical result for stronger correlation functions h(r) we have to look for a simple reaction function $C_2(r)$. In many cases this function will have a very short range, so that we can approximate the convolution term in (19) by

$$\int dr' C_2(r-r')h(r') \sim \int dr' \, \delta(r-r')h(r') = h(r) \; . \tag{31}$$

In the remaining terms C_2 cannot be replaced by δ-functions, this would yield diverging results. We put, turning to spherical coordinates,

$$C_2(r) = \begin{cases} 0 & \text{if} \quad r > a \\ 3/4 \, a^3 & \text{if} \quad r < a \; . \end{cases} \tag{32}$$

In this case the differential equation for h(r) can be solved analytically in both regions. Applying boundary conditions we get a nonlinear complicated algebraic equation for I(N) which can be solved on a computer for each N. The result is that I(N) varies only slightly over N and does not exceed the limits 1 and EB/AC. For I(N) reflects the N-dependence of F_2 one may see that F_2 changes not sensitively with N, as was supposed to decouple the hierarchy of equations for F_2. The solutions shows that varying the range a of C_2 can be compensated by varying D also, although there is no simple relation between both in general.

Now put I(N) into the particle number distribution (17). Varying D over some orders of magnitude, P(N) changes clearly in mean and variance, converging to the homogeneous distribution for $D \to \infty$ or to the other side, for $D \to 0$ to the equilibrium Poisson distribution around the steady state of the linear reaction part.

There is only a certain range where P(N) is really sensitive to D (the diffusion - controlled regime), above that the reaction is practically homogeneous, below, the nonlinear reactions are practically inhibited.

As a conclusion, we may say a generalization of the classical transport theory in diffusion approximation to nonlinear stochastic chemical reactions was proposed. For an ideal system, following a simple nonlinear reaction scheme, analytical solutions for the spatial pair correlation function, as well as for the particle number distribution, have been derived and discussed. When diffusion has extremely high or extremely low values, one of both distributions approaches its equilibrium form, while the other one conserves the nonequilibrium character. It is a plausible result that weaker diffusion inhibits the nonlinear (multi-particle) reaction channels, thus shifting the steady state values in direction to the values of the linear reaction part.

References

1. P. Glansdorff, I. Prigogine: Thermodynamics of Structure, Stability and Fluctuations (Wiley, New York 1971)
2. H. Haken: Synergetics (Teubner, Stuttgart 1973)
3. W. Ebeling, R. Feistel: Physik der Selbstorganisation und Evolution (Akademie-Verlag, Berlin 1982)
4. E.A. Strelcova: Dokl.Akad.Nauk USSR 116, 820 (1957)
5. W. Ebeling: Ann.Physik 16, 147 (1965)
   H. Falkenhagen: Theorie der Elektrolyte (Teubner, Leipzig 1971)
6. M. Malek-Mansour, C. v.d. Broeck, G. Nicolis, J.W. Turner: Ann.Physik 131, 283 (1981)
7. K.J. McNeil, D.F. Walls: J.Stat.Phys. 10, 439 (1974)

Stochastic Thermodynamics

*C. Van den Broeck**

Limburgs Universitair Centrum, B-3610 Diepenbeek, Belgium

1. Introduction

The modern theory of spontaneous fluctuations close to equilibrium was outlined in the classic papers of ONSAGER and MACHLUP /1/. In their theory, the fluctuations are described as stationary, Gaussian continuous Markov processes. The amplitude of the fluctuations (or rather of the random "forces" producing them) is calculated from the Boltzmann relation between the probability $P(\underset{\sim}{x})$ for an equilibrium macrostate $\underset{\sim}{x}$ and the corresponding entropy $S(\underset{\sim}{x})$:

$$P(\underset{\sim}{x}) \sim \exp\left[\frac{S(\underset{\sim}{x})}{k_B}\right] .$$

On the other hand, the theory of discrete Markov processes, based on the so called Master Equation, has been successful in describing the fluctuations in ideal physico-chemical systems in both close to and far from equilibrium states /2/. Moreover, this description relies on the specification of the elementary processes that take place in the system (reaction, diffusion, collision, ...) and does not require the input of thermodynamic information. In fact, we will show how relevant stochastic analogues of nonequilibrium thermodynamic quantities such as entropy production and rate of entropy production can be obtained in this formalism.

As is to be expected, the classic macroscopic thermodynamic expressions for these quantities are recovered in the thermodynamic limit (if the thermodynamic state is unique). However, the interest of a stochastic description is that finite size effects, which include both spontaneous fluctuations and corrections to the macroscopic rate laws, can be studied. The study of fluctuations is a particularly interesting one, since the fluctuations are the intrinsic mechanism by means of which a system tests the stability of its thermodynamic state. Our formalism permits us to identify the contribution of the fluctuations to thermodynamic quantities such as the entropy production and the rate of entropy production. Also, a stochastic version of Prigogine's minimum entropy production theorem can be proven for systems close to equilibrium /3/.

2. The Master Equation

We consider an ideal physico-chemical system described by the discrete state variables $\underset{\sim}{X}$ (e.g. the number of particles with a given position, speed, or of a given species). We suppose that the time evolution of the system can be described by a Markov process. Hence the probability distribution $P(\underset{\sim}{X},t)$ obeys the so called Master Equation /4/:

---
* Bevoegdverklaard Navorser N.F.W.O. België

$$\partial_t P(\underset{\sim}{X},t) = \sum_{\rho} J_{\rho}(\underset{\sim}{X}) \;, \tag{1a}$$

with

$$J_{\rho}(\underset{\sim}{X}) = W_{\rho}(\underset{\sim}{X}-\underset{\sim}{\nu}_{\rho} \rightarrow X)P(\underset{\sim}{X}-\underset{\sim}{\nu}_{\rho},t)-W_{-\rho}(\underset{\sim}{X} \rightarrow \underset{\sim}{X}+\underset{\sim}{\nu}_{\rho})P(\underset{\sim}{X},t)$$

$$= J_{\rho}^{+}(\underset{\sim}{X},t) - J_{\rho}^{-}(\underset{\sim}{X},t) \;, \tag{1b}$$

where the sum over ρ stands for a sum over all the elementary processes that can take place. The quantity W_{ρ} is the probability per unit time for the occurrence of an elementary process ρ. For every such process ρ, which changes the state of the system from X to $\underset{\sim}{X} + \underset{\sim}{\nu}_{\rho}$, there exists an inverse process $-\rho$ with $\underset{\sim}{\nu}_{-\rho} = -\nu_{\rho}$. We have written the Master Equation (1) in such a way to display the quantities $J_{\rho}(\underset{\sim}{X})$, which we will call the stochastic fluxes. For a physicochemical system at equilibrium, the property of detailed balance requires that these fluxes are zero one by one /4/:

$$J_{\rho}^{eq} = 0 \;, \tag{2}$$

whereas at a nonequilibrium steady state, only the sum of these fluxes vanishes:

$$\sum_{\rho} J_{\rho}^{st} = 0 \;.$$

3. Entropy Production of a Markov Process

Our starting point will be the classic definition of the entropy of a Markov process:

$$S(t) = - \sum_{\underset{\sim}{X}} P(\underset{\sim}{X},t) \ln P(\underset{\sim}{X},t) \;. \tag{3}$$

Note that the entropy is, generally speaking, not an increasing function of time, consistent with the fact that we are not dealing with isolated systems.

The time evolution of S can be derived by combining (1) and (3). One obtains /5/:

$$\frac{dS}{dt} = \frac{d_iS}{dt} + \frac{d_eS}{dt} \quad \text{with} \tag{4}$$

$$\frac{d_eS}{dt} = -\frac{1}{2}\sum_{\rho}\sum_{\underset{\sim}{X}} \ln \frac{W_{\rho}(\underset{\sim}{X}-\underset{\sim}{\nu}_{\rho} \rightarrow \underset{\sim}{X})}{W_{-\rho}(\underset{\sim}{X} \rightarrow \underset{\sim}{X}+\underset{\sim}{\nu}_{-\rho})} J_{\rho}(\underset{\sim}{X}) \tag{5}$$

and

$$\frac{d_iS}{dt} = \frac{1}{2}\sum_{\rho}\sum_{\underset{\sim}{X}} J_{\rho}(\underset{\sim}{X},t)\, A_{\rho}(\underset{\sim}{X},t) \tag{6}$$

with (see (2))

$$A_{\rho}(\underset{\sim}{X},t) = \ln \frac{J_{\rho}^{+}(\underset{\sim}{X},t)}{J_{\rho}^{-}(\underset{\sim}{X},t)} \;. \tag{7}$$

Note the analogy of these results with classic nonequilibrium thermodynamics. The stochastic entropy production d_iS/dt is positive semidefinite, $d_iS/dt \geq 0$, whereas the stochastic entropy flux d_eS/dt has no definite sign. We will call $A_{\rho}(\underset{\sim}{X},t)$ the stochastic thermodynamic force. As for the stochastic flux, this quantity is zero at equilibrium:

$$A_{\rho}^{eq} = 0 \;.$$

4. Simple Examples

It is instructive to evaluate the stochastic entropy production for some simple cases.

For the simplest case of a chemical reaction compatible both with an equilibrium state and with the possibility of maintaining non-equilibrium states, e.g.:

$$a \underset{-1}{\overset{1}{\rightleftharpoons}} x \underset{-2}{\overset{2}{\rightleftharpoons}} b \tag{8}$$

with the concentrations of species a and b fixed, one obtains at the steady state /6/:

$$\frac{d_i S}{dt} = (A - X^{st}) \ln \frac{A}{X^{st}} + (B - X^{st}) \ln \frac{B}{X^{st}} \quad , \tag{9}$$

with A and B the number of particles of species a and b and $X^{st} = \frac{A+B}{2}$ the steady state number of particles of species x. For simplicity, we have taken all kinetic constants equal to 1 ($k_1 = k_{-1} = k_2 = k_{-2} = 1$). We note that the above result (9) is identical to the classic Gibbs entropy production, $d_i S/dt = d_i S/dt|_G$. Also, one finds the classic Gibbs result for the stochastic entropy production associated to a diffusion process being modelled by a random walk. These processes are linear ones. However, for the nonlinear chemical reaction scheme

$$a + x \underset{-1}{\overset{1}{\rightleftharpoons}} 2x \qquad\qquad x \underset{-2}{\overset{2}{\rightleftharpoons}} b \ , \tag{10}$$

one obtains (at the steady state, $\delta X = X - \langle X \rangle^{st}$)

$$\frac{d_i S}{dt} = \frac{d_i S}{dt}\bigg|_G - \frac{1}{A} \left[\langle \delta X^2 \rangle^{st} - \langle X \rangle^{st} \right] \ln \frac{A}{X^{st}} \ . \tag{11}$$

The second term in the r.h.s. of (11) is a finite size correction (it is of order 1, whereas the Gibbs entropy production is extensive). We shall call it the entropy production of the fluctuations and write $d_i S/dt|_F$. It is zero at the equilibrium state, since the latter is characterized by Poissonian fluctuations, $\langle \delta X^2 \rangle^{eq} = \langle X \rangle^{eq}$. For the above-considered model, $d_i S/dt|_F < 0$ in a nonequilibrium steady state. This turns out to be a general feature for chemical systems close to equilibrium, but no definite results can be obtained for far-from-equilibrium states.

5. Minimum Entropy Production Theorem for Discrete Markov Processes

In full analogy with the classic thermodynamic theory, one obtains for the rate of the stochastic entropy production:

$$\frac{d}{dt}\left(\frac{d_i S}{dt}\right) = \frac{d_A}{dt}\left(\frac{d_i S}{dt}\right) + \frac{d_J}{dt}\left(\frac{d_i S}{dt}\right) \tag{12}$$

with

$$\frac{d_A}{dt}\left(\frac{d_i S}{dt}\right) = \frac{1}{2} \sum_{\rho} \sum_{\underline{X}} J_\rho(\underline{X},t) \frac{\partial A_\rho(\underline{X},t)}{\partial t} = - \sum_{\underline{X}} \frac{[\partial_t P(\underline{X},t)]^2}{P(\underline{X},t)} \leq 0 \tag{13}$$

and

$$\frac{d_J}{dt}\left(\frac{d_i S}{dt}\right) = \frac{1}{2}\sum_{\varrho}\sum_{\underset{\sim}{X}} \frac{\partial J_{\varrho}(\underset{\sim}{X},t)}{\partial t} A_{\varrho}(\underset{\sim}{X},t) . \tag{14}$$

The result (13) is reminiscent of the general evolution criterion derived by GLANSDORFF and PRIGOGINE /7/. In order to derive the minimum entropy production theorem one has to specify, in addition, that the system is close to equilibrium. We express this in the following way:

$$W_{\varrho} = W_{\varrho}^{eq} + \delta W_{\varrho} , \tag{15}$$

$$P(\underset{\sim}{X},t) = P^{eq}(\underset{\sim}{X}) + \delta P(\underset{\sim}{X},t) , \tag{16}$$

being understood that quantities of order δ^2 are to be neglected. The equilibrium transition probabilities W_{ϱ}^{eq} and equilibrium distribution P^{eq} must satisfy the detailed balance condition (2). From (1b), (7), (15) and (16) one concludes:

$$A_{\varrho}(\underset{\sim}{X},t) = \frac{J_{\varrho}(\underset{\sim}{X},t)}{W_{-\varrho}^{eq}(\underset{\sim}{X} \to \underset{\sim}{X} + \underset{\sim}{\nu}_{-\varrho})\, P^{eq}(\underset{\sim}{X})} . \tag{17}$$

From this linear relation between fluxes and forces, it follows:

$$\frac{d_A}{dt}\left(\frac{d_i S}{dt}\right) = \frac{d_J}{dt}\left(\frac{d_i S}{dt}\right) , \tag{18}$$

hence

$$\frac{d}{dt}\left(\frac{d_i S}{dt}\right) = -2\sum_{\underset{\sim}{X}} \frac{[\partial_t P(\underset{\sim}{X},t)]^2}{P(\underset{\sim}{X},t)} \leqslant 0 . \tag{19}$$

Since $d_i S/dt$ is positive semi-definite, we conclude that the stochastic entropy production is a Ljapunov functional in the linear region of irreversible thermodynamics. Thus, the stationary distribution is characterized by minimum entropy production.

6. Discussion

The stability of systems described by a Master Equation has been related by previous authors to the properties of the so called H-functional /8/. However, the decrease of this H-functional in time is satisfied identically, independent of the stability and the distance from equilibrium. For this reason the thermodynamic interpretation of the H-functional is incontrovertible. Macroscopic stability in connection with some properties of the fluctuations has also been discussed by KEIZER /9/. The central quantity in his approach is the variance of the fluctuations whose thermodynamic interpretation is not obvious.

We hope that our thermodynamic-minded approach opens a new way to the familiar and yet elusive question of the effect of fluctuations on stability.

References

1. L. Onsager and S. Machlup: Phys.Rev. 91. 1505 (1953); 91, 1512 (1953)
2. See e.g. G. Nicolis and I. Prigogine: Selforganization in Non-equilibrium systems (Wiley, New York 1977)
3. A more detailed account of this theory can be found in
   a) Luo Jiu-li, C. Van den Broeck and G. Nicolis: Z.Phys.

B56, 165 (1984)
b) Luo Jiu-li: Ph.D. dissertation, University of Brussels (1984)
4. See e.g. N.G. Van Kampen: Stochastic Processes in Physics and Chemistry (North-Holland, Amsterdam 1982)
5. J. Schnakenberg: Rev.Mod.Phys. 48, 571 (1976)
6. For more details, see reference 3b
7. P. Glansdorff and I. Prigogine: Thermodynamic Theory of Structure, Stability and Fluctuations (Wiley, London 1971)
8. F. Schlögl: Z.Phys. 243, 303 (1971); Z.Phys. 248, 446 (1972)
C.W. Gardiner: J.Chem.Phys. 70, 5778 (1979)
9. J. Keizer: J.Chem.Phys. 69, 2609 (1978); Acc.Chem.Res. 12, 243 (1979)

Stochastic Measures as Critical Quantities in Nonequilibrium Phase Transitions

F. Schlögl

RWTH Aachen, Institut für Theoretische Physik, Templergraben 64,
D-5100 Aachen, Fed. Rep. of Germany

1. Introduction

As already discussed in earlier papers of the author /1/, the bit-number cumulants of a probability distribution over the microstates of a system are sensitive to interior correlations between parts of the system. The second of these cumulants, the bit-number variance, is sensitive in a higher degree than the first one, the informational entropy /2/. Of special interest are the bit-number cumulants of the probability distributions in Gibbs phase space of the microstates described by the canonical variables of Hamilton mechanics. Equilibrium entropy is a first cumulant and specific heat is a second cumulant of this type. The sensitivity of the bit-number variance is the reason for the dramatic critical behaviour of specific heat in an equilibrium phase transition.

2. Critical Quantities in Equilibrium and in Nonequilibrium Phase Transitions

It was pointed out by several authors /3-7/ that in certain physical systems of different kinds, cooperative effects can be observed which, notwithstanding that they occur far away from any thermal equilibrium, show properties closely analogous to those of equilibrium phase transitions. We then speak of "nonequilibrium phase transitions". In the following we will deal with such transitions which can be observed in the diagram of steady (stationary) states of an open system. (We do not speak of structure in time.) In case of a nonequilibrium phase transition, the steady state diagram, in particular with respect to singularities, looks like an equilibrium state diagram of a system with a phase transition. This equilibrium state diagram is given by the "thermal state equation". By this the critical behaviour of certain quantities is already determined, like the singularity of the order parameter and of certain susceptibilities (so we call derivatives of a thermal quantity with respect to another one). Yet not determined by the thermal state equation is the behaviour of specific heat. For this we need the caloric state equation as an independent information. We can obtain all these informations by the knowledge of the energy as function of thermal variables.

In nonequilibrium we do not know a quantity which plays an analogous role like free energy in equilibria. I mean a thermodynamic function which distinguishes steady states far away from any equilibrium by its extrema and which gives an analogous additional information like the caloric equation for equilibria (the principle that steady states are distinguished by minimum entropy production is valid only in linear thermodynamics near an equilibrium).

Specific heat is a special susceptibility. It is the derivative of energy E with respect to temperature T:

$$c = \partial E / \partial T \ . \qquad (2.1)$$

Temperature is the thermal adjoint to energy. Energy is the generator of time-shift against which an equilibrium is invariant, unlike steady nonequilibrium states which belong to an environment changing with time. Temperature is distinguished in equilibrium phase transitions as a natural control parameter. This distinction does no more exist in nonequilibria. Specific heat is also proportional to the variance of energy in a canonical ensemble:

$$c \sim \langle (\Delta E)^2 \rangle \ . \qquad (2.2)$$

And this provides a key to find an analogous quantity in nonequilibria.

3. Bit-Number Cumulants

We consider a sample set of events which are uniquely characterized by a set

$$\xi = (\xi_1, \xi_2, \ldots) \qquad (3.1)$$

of continuous parameters ξ_s. A special case which will be considered later is that these events are microstates of a physical system in Gibbs phase space. Then the set ξ is the set of the canonical variables q_ι and their adjoint momenta p_ι. Let $\varrho(\xi)$ be the probability density of the events in ξ-space. We call

$$b(\xi) = -\log \varrho(\xi) \qquad (3.2)$$

the corresponding bit-number because it is related to the number of bits which are necessary to select the event ξ if $\varrho(\xi)$ is already known. The mean value of

$$\langle b \rangle = \int d\xi \, \varrho \, b \qquad (3.3)$$

is Shannon's informational entropy, a fundamental measure of the distribution ϱ in information theory /8/.

This mean value of b is the first cumulant of b. We can define bit-number cumulants Γ_k of any order k by the expansion

$$\log \langle e^{\alpha b} \rangle = \sum_{k=1}^{\infty} \alpha^k \Gamma_k / k! \ . \qquad (3.4)$$

These stochastic measures Γ_k are distinguished by the following two features:

(a) They are additive for uncorrelated subsystems.
(b) They are invariant with respect to volume-conserving transformations of ξ-space.

The feature (a) allows us to use the difference

$$\Gamma_k - \Gamma_k^{I} - \Gamma_k^{II} \qquad (3.5)$$

as an ordered system of measures for the correlations between two subsystems I, II of which the system is composed. The feature (b) makes Γ_k invariant against Liouville motion in Gibbs phase space.

The following statement shows that the second cumulant Γ_2 is more sensitive to the onset of correlations than the first one, the information entropy Γ_1. We consider a given distribution $\varrho_{i_1 i_2 \ldots}$ of a system which is composed of subsystems the states of which are $i_1, i_2, \ldots$ The probability distribution of the first subsystem is the marginal distribution

$$\varrho_{i_1} = \sum_{i_2 i_3 \dots} \varrho_{i_1 i_2 \dots} \,. \tag{3.6}$$

In an analogous way we define marginal distributions ϱ_{i_s} for any subsystem. The product of all of them

$$\varrho^{\circ}_{i_1 i_2 \dots} = \varrho_{i_1} \varrho_{i_2} \dots \tag{3.7}$$

is the distribution in which the probability densities ϱ_{i_s} for the subsystems are unchanged but all correlations between them are suppressed. We call it the uncorrelated distribution to $\varrho_{i_1 i_2 \dots}$.
We can always write

$$\varrho_{i_1 i_2 \dots} = (1 + \varepsilon_{i_1 i_2 \dots}) \varrho^{\circ}_{i_1 i_2 \dots} \,, \tag{3.8}$$

where ε describes the correlations in a given microstate. For small ε, say for the onset of correlations it turns out that Γ_2 depends on ε in first order already, whereas Γ_1 depends but in second order.

4. Bit-Number Variance as Thermodynamic Quantity

A macroscopic thermal equilibrium state can be described by a set of variables M^ν which in statistical theory are mean values of phase functions $M^\nu(\xi)$ in Gibbs phase space ξ. The adequate probability distribution in ξ-space belonging to maximum entropy <b> is

$$\underset{\sim}{\varrho} = \exp[\Phi - \lambda_\nu \underset{\sim}{M}^\nu] \,. \tag{4.1}$$

The underlining tilda denotes the phase functions. Greek sub- and superscripts occurring twice in a product are summation dummies. As a rule M^ν are extensive and λ_ν intensive variables describing the same macrostate. The construction of $\underset{\sim}{\varrho}$ in the form (4.1) is interpreted by JAYNES /9/ as the method of "unbiased guess". We call this $\underset{\sim}{\varrho}$ a generalized canonical ensemble. It is well known that the same macrostate can be described alternatively by a microcanonical, canonical, or a grand canonical ensemble. This means that we can use different $\underset{\sim}{\varrho}$ with less than the whole set M^ν for the same macrostate. So we have generally to distinguish between M^ν entering into $\underset{\sim}{\varrho}$ as phase functions, this is as random quantities, and those M^ν entering as fixed parameters. Correspondigly Γ_2 becomes different for the different $\underset{\sim}{\varrho}$. The specific heat is such a Γ_2 and becomes thus different for let us say constant volume than for constant pressure.

It is found by examples that this kind of specific heat shows the most dramatic critical behaviour in equilibrium phase transitions, which is Γ_2 to a $\underset{\sim}{\varrho}$ in which the order parameter is a random variable M^ν.

Now we have to ask for an adequate probability distribution describing a given macroscopic nonequilibrium state with local and time-dependent macroscopic variables $M^\nu(\underline{x},t)$. These variables then are densities in three-dimensional geometric space $\underline{x}$. In the so-called Schrödinger picture these variables are interpreted as mean values of phase space functions $\underset{\sim}{M}^\nu(\underline{x};\xi)$ in a time-dependent probability distribution $\underset{\sim}{\varrho}(t;\xi)$. If we apply Jaynes method of unbiased guess, starting with the instantaneous mean values $M^\nu(\underline{x},t)$ we obtain the so-called "local equilibrium distribution"

$$\underset{\sim}{\varrho}^L(t) = \exp[\Phi(t) - \lambda(t).\underset{\sim}{M}] \,, \tag{4.2}$$

where we use the symbolic abbreviation

$$\lambda . M = \int d^3x \, \lambda_\nu(\underline{x}) M^\nu(\underline{x}) \quad . \qquad (4.3)$$

This $\varrho^L(t)$ is, however, deficient because no knowledge about the dynamics is included in the construction. Therefore this $\varrho^L(t)$ is not able to describe correlations and transport phenomena correctly.

The construction of a suitable distribution ϱ belonging to the nonequilibrium state was first proposed by H. MORI /10/, which indeed has given good results in transport theory and similar questions. This construction implies that the time scale of the microscopic processes is distinctly separated from that of macroscopic processes. This is a standard condition for the applicability of a thermodynamic description. If it is fulfilled, then there exists a large plateau of intermediate times τ . In such a time τ the system is practically isolated, and therefore its dynamics in Gibbs phase space is a Liouville motion. ϱ at time t is constructed by the application of this dynamics during a time τ on the local equilibrium distribution ϱ^L at the earlier time $t-\tau$. By use of the Liouville operator $\mathcal{L}$ this is

$$\varrho(t) = e^{-i\tau\mathcal{L}} \varrho^L(t-\tau) \quad . \qquad (4.4)$$

The result of this construction is identical with expressions based on different arguments later by McLENNAN /11/ and ZUBAREV /12/. Due to the supposed time-scale separation the result is independent of the choice of τ if only on the plateau. ϱ belongs practically to the same mean values $M^\nu(\underline{x},t)$ as ϱ^L, yet the two distributions are very different. In ϱ all interior correlations are fully developed which were suppressed in ϱ^L . Now the invariance property (b) mentioned in the preceding chapter becomes important. Due to it, all bit-number cumulants Γ_k for ϱ are the same as for ϱ^L at the earlier time. They are functionals of the variables λ_ν, M^ν and therefore practically the same at time t and $t-\tau$. This means, due to the time-scale separation, the Γ_k are the same for ϱ as for ϱ^L. This is an important and nontrivial justification of the local equilibrium approximation for all Γ_k. Thus the bit-number variance for the Mori distribution is

$$\Gamma_2 = - \int d^3x d^3x' \, \lambda_\mu(\underline{x}) \lambda_\nu(\underline{x}') \, \delta M^\mu(\underline{x}) / \delta \lambda_\nu(\underline{x}') \quad . \qquad (4.5)$$

The derivative in this expression is a functional derivative in space. In macroscopic systems composed of different homogeneous parts, as a rule the correlation lengths are so small that these derivatives, with the exception that correlations going through the interfaces between the parts, lead to derivatives of variables of one part with respect to variables of the neighboured part.

For systems with a nonequilibrium phase transition Γ_2 formed with the distribution in which the order parameter is a random quantity is a distinguished candidate for a characteristic critical behaviour. Several examples have been discussed in ref. /2/ (see also /13-17/).

References

1. F. Schlögl: Z.Physik 267, 77 (1974); Z.Physik B 20, 177 (1975); 22, 301 (1975); Phys.Rep. 62 (No.4), 267 (1980)
2. F. Schlögl: Z.Physik B 52, 51 (1983)
3. H. Haken, W. Weidlich: Quantum Theory of the Laser, Rendiconti Scuola "E. Fermi" XLII
4. R. Graham, H. Haken: Z.Physik 213, 420 (1968)
5. E. Pytte, H. Thomas: Phys.Rev. 179, 431 (1969)

6. A.F. Volkov, M. Kogan: Sov.Phys.Usp. 11, 881 (1969)
7. V. de Giorgio, M.O. Scully: Phys.Rev. A 2, 1170 (1970)
8. C. Shannon, W. Weaver: The Mathematical Theory of Communication (University of Illinois Press, Urbana 1949)
9. E.T. Jaynes: Phys.Rev. 106, 620 (1957)
10. H. Mori: Phys.Rev. 115, 298 (1959)
11. J.A. McLennan: Advances in Chemical Physics (Ed.: I. Prigogine) (Interscience, New York 1963)
12. D.N. Zubarev: Sov.Phys.Dokl. 10, 526 (1965)
13. F. Schlögl: Z.Physik 253, 147 (1972)
14. H. Busch: Annalen d. Physik 64, 401 (1921)
15. B. Ross, J. Litster: Phys.Rev. A 15, 1246 (1977)
16. D. Bedeaux, P. Mazur, R.A. Pasmanter: Physica 86A, 355 (1977)
17. H. Haken: Synergetics, an Introduction (Springer, Berlin Heidelberg New York 1977)

On the Kinetics of Nucleation in Isochoric Gases

L. Schimansky-Geier[1], *F. Schweitzer*[2], *W. Ebeling*[1], *and H. Ulbricht*[2]

[1]Humboldt-Universität Berlin, Sektion Physik, Bereich 04, PSF 1297, DDR-1086 Berlin, GDR

[2]Wilhelm-Pieck-Universität Rostock, Sektion Physik, Universitätsplatz 3, DDR-2500 Rostock, GDR

1. Introduction

The process of phase separation in gases has been studied in terms of several different theoretical models (for review see refs. /1-3/). The classical nucleation theory /4-7/ is based on the droplet model, corresponding to finite and discrete fluctuations. Other approaches use theoretical methods with a continous description of the density fluctuations /8-13/. The classical nucleation theory /4-7/ is based on the assumption, that the density of free particles is constant and that the number of all particles in the system is not limited.

These assumptions correspond to a nucleation process in an infinite system. In the present work the problem will be considered how the finite system size will modify the nucleation process. Basing on the droplet model we propose the following model:
We consider a gas of N particles in a closed system with a constant volume V at a temperature T; it means:

$$N = \text{const.};\ V = \text{const.};\ T = \text{const.} \tag{1.1}$$

Due to interactions between the particles (e.g. chemical reactions, collisions) the N particles will bond in clusters and a distribution of N_1 monomers (free particles), N_2 dimers (bound states of 2 particles), ..., N_N N-mers develops. This distribution will be described by

$$\underline{N} = \{N_1\ N_2\ \ldots N_l\ \ldots N_N\}\ . \tag{1.2}$$

Because of (1.1), the relation (1.3) holds:

$$N = \sum_{l=1}^{N} l\, N_l = \text{const.} \tag{1.3}$$

l is the number of particles in the cluster; the number of clusters consisting of l particles is denoted by N_l.

From (1.3) it follows further that the maximal number of N_l is given by

$$0 \leq N_l \leq \frac{N}{l} \quad (l=1,\ldots,N)\ . \tag{1.4}$$

The clusters are assumed to be an ideal mixture.

2. Equilibrium Distribution

With respect to the constraints (1.1) the equilibrium distribution of clusters is given by the canonical ensemble. We denote the probability distribution for the N particles in the phase space as $w(q_1 \ldots q_N, p_1 \ldots p_N)$, where q_l are the space coordinates and p_l are the momentum coordinates of the particles with the number $l=1,\ldots,N$. In thermodynamic equilibrium the following relation is valid /14/:

$$w^o(q_1 \ldots p_N) = \frac{1}{Z} \exp\left\{-\frac{1}{k_B T} H(q_1 \ldots p_N)\right\} \quad . \tag{2.1}$$

$H(q_1 \ldots p_N)$ denotes the Hamiltonian of the N particle system, Z is the canonical partition function of the N particles in an atomic picture:

$$Z = \exp\left\{-\frac{1}{k_B T} F(T,V,N)\right\} \quad . \tag{2.2}$$

$F(T,V,N)$ is the free energy of the N particles in the volume V with full interactions.

We consider now a certain configuration $C(\underline{N}) = C(N_1 \ldots N_N)$ of the N particles that corresponds to the assumed cluster distribution (1.2). We define the probability of the equilibrium distribution for this configuration by an integral of $w(q_1 \ldots p_N)$ over the subspace $C(\underline{N})$:

$$P^o(\underline{N}) = \int_{C(\underline{N})} w^o(q_1 \ldots p_N) dq_1 \ldots dp_N \quad . \tag{2.3}$$

With (2.1) and (2.2) we get

$$P^o(\underline{N}) = \exp\left\{\frac{1}{k_B T} F(T,V,N)\right\} \int_{C(\underline{N})} \exp\left\{-\frac{1}{k_B T} H(q_1 \ldots p_N)\right\} dq_1 \ldots dp_N \quad . \tag{2.4}$$

The partition function of the assumed cluster distribution (1.2) can be calculated by /15/:

$$Z(T,V,\underline{N}) = \int_{C(\underline{N})} \exp\left\{-\frac{1}{k_B T} H(q_1 \ldots p_N)\right\} dq_1 \ldots dp_N \quad . \tag{2.5}$$

For a configuration $C(\underline{N})$ which in section 3 will be assumed as an ideal mixture of the clusters with the constraint (1.3) the free energy is introduced as follows:

$$F(T,V,\underline{N}) = -k_B T \ln Z(T,V,\underline{N}) \quad . \tag{2.6}$$

(Note, that $F(T,V,\underline{N})$ (2.6) and $F(T,V,N)$ (2.2) are different functions.)

In this way we get the correct probability $P^o(\underline{N})$ of the equilibrium distribution:

$$P^o(N_1 N_2 \ldots N_N) = \exp\left\{\frac{1}{k_B T}\left[F(T,V,N) - F(T,V,N_1 N_2 \ldots N_N)\right]\right\} \quad . \tag{2.7}$$

3. The Free Energy

The free energy (2.6) of the ideal mixture of clusters with the distribution (1.2) is determined by the partition function (2.5). With the correct normalization it results:

$$Z(T,V,\underline{N}) = \prod_{l} \frac{1}{h^{3N_l} N_l!} \int \exp\left\{-\frac{1}{k_B T} H_l\right\} dq_1^{N_1} \ldots dp_N^{N_N} . \qquad (3.1)$$

H_l is the Hamiltonian of the cluster with the particle number l and the mass

$$M_l = \frac{M}{N_A} \, l ,$$

(M - molar mass, N_A - Avogadro-constant)

$$H_l = \frac{1}{2M_l} (p_{1l}^2 + p_{2l}^2 + p_{3l}^2) + f_l , \qquad (3.2)$$

f_l denotes a potential contribution which depends only on the size l of the cluster.

Integration of (3.1) and use of (2.6) leads to the following expression for the free energy:

$$F(T,V,\underline{N}) = \sum_{l=1}^{N} N_l \left\{ f_l + k_B T \left(\ln \frac{N_l}{V} \lambda_l^3 - 1\right)\right\} , \qquad (3.3)$$

with

$$\lambda_l = h \left[2\pi M_l \, k_B T\right]^{-1/2} . \qquad (3.4)$$

To determine the potential term f_l we choose a first approximation similar to the theory of atomic nuclei which includes only volume and surface effects:

$$f_l(T) = - Al + Bl^{2/3} . \qquad (3.5)$$

The second term of eq. (3.5) is proportional to the surface area and the surface tension σ. It follows

$$B = 4\pi \left(\frac{4\pi}{3} c N_A\right)^{-2/3} \sigma , \qquad (3.6)$$

where c [mol/m^3] is the molar density in the cluster. The boundary conditions for the potential part are given by

$$f_1 = 0 ,$$
$$f_2 = \varphi_2 = \text{two-particle binding energy} . \qquad (3.7)$$

The free energy F = G-pV given by (3.3) includes the contribution of the partial pressures

$$- pV = - \sum_{l=1}^{N} p_l V = - \sum_{l=1}^{N} N_l k_B T \qquad (3.8)$$

and the contribution of the Gibbs potential

$$G = \sum_{l=1}^{N} N_l \, \mu_l \qquad (3.9)$$

with the chemical potential of a cluster with size l

$$\mu_l = f_l(T) + k_B T \ln \frac{N_l}{V} \lambda_l^3 . \qquad (3.10)$$

4. The Master Equation

To discuss the time evolution of the cluster size distribution we suppose the following assumptions:
(i) The growth and shrink of a cluster is due only to an addition or evaporation of monomers (Fig. 1). In terms of chemical kinetics this process can be represented by:

$$A_l + A_1 \underset{w^-}{\overset{w^+}{\rightleftharpoons}} A_{l+1} , \qquad (4.1)$$

where w^+ and w^- are the probabilities per unit time of the reaction in the given directions.
(ii) Interactions between the clusters with $l \geqslant 2$, like coagulation processes or collisions between two or more clusters are not taken into account. The probabilities of these events should be negligible in comparison with the probabilities of the reaction (4.1).

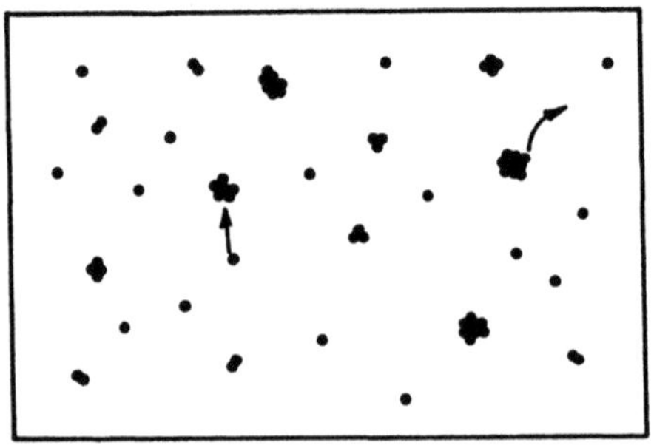

Fig. 1 Typical cluster distribution in a finite volume V and the elementary processes of attachment and evaporation of monomers

The interactions between the clusters are taken into account via their influence on the pressure in the system. This pressure is given as a sum of the partial pressures of the different kinds of clusters (3.8). We define the probability

$$P(N_1 N_2 \dots N_N, t) = P(\underline{N}, t) \qquad (4.2)$$

that at a given time t there exist N_1 monomers, N_2 dimers ... etc. in the system. Then the time evolution of the cluster distribution (1.2) corresponds to the change of $P(\underline{N},t)$ with time. This evolution can be described by a Master equation

$$\frac{\partial P(\underline{N},t)}{\partial t} = \sum_{\underline{N}'} \{ w(\underline{N}|\underline{N}') P(\underline{N}',t) - w(\underline{N}'|\underline{N}) P(\underline{N},t) \} . \qquad (4.3)$$

$\underline{N}'$ specifies those cluster distribution which are attainable from the assumed distribution $\underline{N}$ via the reaction (4.1). The quantities $w(\underline{N}|\underline{N}')$ are the transition probabilities per time unit for the transition from $\underline{N}'$ to $\underline{N}$.

Equation (4.3) allows us to introduce a probability flux

$$J(\underline{N}|\underline{N}') = w(\underline{N}|\underline{N}')P(\underline{N}',t) - w(\underline{N}'|\underline{N})P(\underline{N};t) \ . \tag{4.4}$$

Substitution of (4.4) into (4.3) results in

$$\frac{\partial P(\underline{N},t)}{\partial t} = \sum_{\underline{N}'} J(\underline{N}|\underline{N}') \ . \tag{4.5}$$

The stationary solution of (4.5) implies that

$$\sum_{\underline{N}'} J(\underline{N}|\underline{N}') = 0 \ . \tag{4.6}$$

This is the so-called KIRCHHOFF solution /16/. With the condition of detailed balance follows a more restrictive form of (4.6):

$$J(\underline{N}|\underline{N}') = 0. \tag{4.7}$$

This consistency condition (4.7) means:

$$w(\underline{N}|\underline{N}')P^{o}(\underline{N}') = w(\underline{N}'|\underline{N})P^{o}(\underline{N}) \ . \tag{4.8}$$

In this equation $P^{o}(\underline{N})$ is the probability of the equilibrium distribution (2.7). To construct the transition probabilities my means of (4.8) we need one kinetic assumption with respect to (4.1). In agreement with earlier investigations /17/ we assume that the probability of an attachement of a monomer to a cluster of the size l increases with the surface of the cluster and with the density of clusters of size l and the density of monomers. In this way we get for the special process of attachment: $l > 2$

$$W(N_1-1 \ldots N_l-1 N_{l+1}+1 \ldots N_N | N_1 \ldots N_l N_{l+1} \ldots N_N)$$
$$\equiv w_l^{+}(N_1 N_l) = \alpha \, l^{2/3} \, N_l \, \frac{N_1}{V} \ . \tag{4.9}$$

We would like to underline that the probabilities of the growth process of clusters of different sizes l=1...N are correlated since the number of particles is conserved (1.3).

Note, that for l=1 the number of monomers changes from N_1 to N_1-2 by creating a dimer. The transition probability in this case reads:

$$w_1^{+}(N_1-2 \ N_2+1 \ N_3 \ldots N_N | N_1 N_2 N_3 \ldots N_N) \equiv w_1^{+}(N_1) = \alpha_1 \frac{N_1(N_1-1)}{V} \ . \tag{4.10}$$

From (4.8) with (2.7) and (4.9) we get the transition probability for the oppositve process of (4.9) (evaporation) /18/:

$$w(N_1 \ldots N_l N_{l+1} \ldots N_N | N_1-1 \ldots N_l-1 N_{l+1}+1 \ldots N_N)$$
$$\equiv w_{l+1}^{-}(N_{l+1}+1) = \alpha \, l^{2/3} N_l \, \frac{N_1}{V} \exp\left\{ \frac{1}{k_B T} \right.$$
$$\left. * \left[F(T,V,N_1-1 \ldots N_l-1 \ N_{l+1}+1 \ldots N_N) - F(T,V,N_1 \ldots N_l \ N_{l+1} \ldots N_N)\right] \right\} . \tag{4.11}$$

If we neglect in (4.11) a small term $(l/l+1)^{13/6}$, which is nearly equal to one, we get for the transition probability for the evaporation process $N_1 \ldots N_l \ldots N_N \longrightarrow N_1+1 \ldots N_{l-1}+1 \ N_l-1 \ldots N_N$:

$$w_l^{-}(N_l) = \alpha \, l^{2/3} \, N_l \, \frac{1}{\lambda_1^3} \exp \frac{1}{k_B T}(f_l - f_{l-1}), \tag{4.12}$$

if $l = 2$, $\alpha \rightarrow \alpha_1$.

The transition probability of the evaporation of a monomer is not determined by the whole cluster distribution as in (4.9). Again, this transition probability is modified for the evaporation of a dimer, because the transition changes N_1, N_2 to (N_1+2), (N_2-1). We underline that in our model the transition probabilities for all processes different from (4.9) and (4.12) are assumed to be zero.

With the introduced transition probabilities the master equation (4.3) reads in an explicit form with $W_N = W_{N+1} = 0$:

$$\dot{P}(\underline{N},t) = w_2^-(N_2+1)P(N_1-2N_2+1\ldots N_N,t)-w_1^+(N_1)P(\underline{N},t)+w_3^-(N_3+1)$$

$$\cdot P(N_1-1,N_2-1N_3+1\ldots N_N,t)+w_1^+(N_1+2)\; P(N_1+2N_2-1\ldots N_N,t)$$

$$-[w_2^+(N_1N_1)-w_2^-(N_2)]\,P(\underline{N},t)+\sum_{l=3}^{N} w_{l+1}^-(N_{l+1}+1)P(N_1-1\ldots N_{l-1}-1N_{l+1}+1\ldots N_N,t)$$

$$+w_{l-1}^+(N_1+1N_{l-1}+1)P(N_1+1\ldots N_{l-1}+1N_l-1\ldots N_N,t)$$

$$-[w_l^+(N_1N_l)+w_l^-(N_l)]\,P(\underline{N},t)\;. \tag{4.13}$$

5. Discussion

(i) First we investigate the stationary probability distribution $P^0(N_1N_2\ldots N_l\ldots N_N)$ given by (2.7). The extremal probability we find from $\partial P^0/\partial N_l=0$. It yields the equilibrium distribution of clusters:

$$\frac{N_l^0\lambda_l^3}{V} \approx \exp\left\{l\;\ln\frac{N_1^0\lambda_1^3}{V}-\frac{f_l}{k_BT}\right\}\,,\qquad l=2,\ldots,N \tag{5.1}$$

where $N_1^0 = N - \sum_{l=2}^{N} l\,N_l^0$.

Due to this boundary condition the solution of (5.1) is complicated. Inside the binodal region we generally expect multiple solutions in the $(N_1\ldots N_N)$ space. At least the system is a bistable one where two stable cluster distributions can exist separated by regions of a minimal probability. Figure 2 demonstrates that we find one regime with a large number of monomers and a second regime with a stable configuration of large clusters in coexistence with a small number of monomers.

The question which cluster distribution is realized depends on the absolute value of the probability distribution in the $(N_1\ldots N_N)$ space, which is determined by the thermodynamic parameters N,V,T. The problem of transitions between the two stable states of the cluster distribution can be turned out with the concept of the mean first passage time /19/ which allows the calculation of the mean transition time from the monomer phase to the cluster phase.

It can be estimated by the consideration of the eigenvalue problem of (2.7), by finding the smallest eigenvalue χ_1 which is different from zero. In bistable systems χ_1 is separated from the spectrum of the other eigenvalues $\chi_i(i=2,3,\ldots)$ /20/.

(ii) Let us derive the equations of the mean values /21/. The mean cluster number with size k

$$\langle N_k(t)\rangle = \sum_{\{N_i\}} N_k\; P(N_1\ldots N_N,t)$$

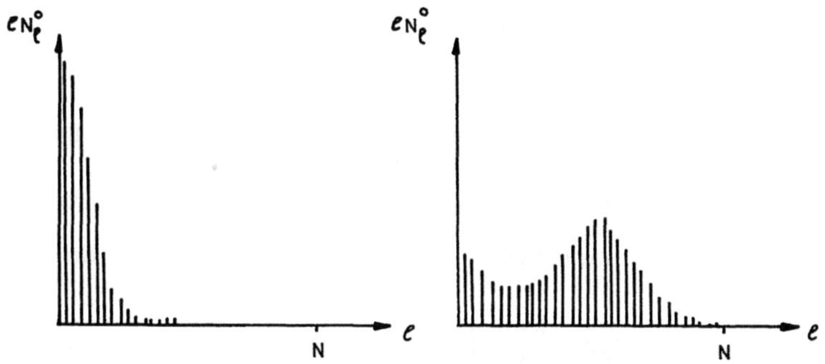

Fig. 2 Sketch of the equilibrium cluster distributions:
a) corresponds to a monomer phase, in b) the cluster phase is dominating

obeys the system of equations

$$\frac{\partial}{\partial t}\langle N_l\rangle = \langle \frac{\partial}{\partial l}(w^-_{l+1}(N_{l+1}) - w^+_l(N_1 N_k))\rangle \quad , \tag{5.2}$$

$$\frac{\partial}{\partial t}\langle N_1\rangle = -\langle w^+_1(N_1) - w^-_2(N_2) + \sum_{l=1}^{N-1}(w^+_l(N_1 N_l) - w^-_{l+1}(N_{l+1}))\rangle . \tag{5.3}$$

The derivative in (5.2), which is a discrete one, expresses the conversion of the cluster number during the growth and shrinkage of clusters ($l > 2$). The mean cluster number changes only due to the formation or evaporation of dimers

$$\frac{\partial}{\partial t}\sum_{l=2}^{N}\langle N_l\rangle = \langle w^+_1(N_1) - w^-_2(N_2)\rangle . \tag{5.4}$$

The alteration of the actual pressure is defined by the equation for the mean numbers of clusters and monomers:

$$\dot{P}(t) = \frac{k_B T}{V}\frac{\partial}{\partial t}\sum_{l=1}^{N}\langle N_l\rangle = -\frac{k_B T}{V}\sum_{l=1}^{N-1}\langle(w^+_l(N_1 N_l) - w^-_{l+1}(N_{l+1}))\rangle . \tag{5.5}$$

Obviously it follows for the whole number of particles in the considered case:

$$\frac{\partial}{\partial t}\left\{\langle N_1\rangle + \sum_{l=2}^{N}\langle l\, N_l\rangle\right\} = 0 \quad . \tag{5.6}$$

(iii) Treating N_l as a continuous function of l, we make use of a Kramers-Moyal expansion /22/ in (5.2). Approximating $\langle N_1\, N_l\rangle \approx \langle N_1\rangle\langle N_l\rangle$ we obtain a Fokker-Planck equation for the mean cluster distribution by neglecting higher than the second derivative:

$$\frac{\partial}{\partial t}\langle N_l\rangle = -\frac{\partial}{\partial l}\left\{\langle N_l\rangle \,\alpha\, l^{2/3}\left\{\frac{\langle N_1\rangle}{V} - \frac{1}{\lambda_1^3}\exp\left(\frac{1}{k_B T}\frac{\partial f_l}{\partial l}\right)\right\}\right\}$$
$$+\frac{1}{2}\frac{\partial^2}{\partial l^2}\left\{\langle N_l\rangle \,\alpha\, l^{2/3}\left\{\frac{\langle N_1\rangle}{V} + \frac{1}{\lambda_1^3}\exp\left(\frac{1}{k_B T}\frac{\partial f_l}{\partial l}\right)\right\}\right\} + \ldots \qquad l \geq 2 \tag{5.7}$$

If we further neglect the diffusion due to fluctuations we derive the deterministic case (Liouville equation)

$$\frac{\partial}{\partial t}\langle N_l\rangle = -\frac{\partial}{\partial l}\,\alpha l^{2/3}\langle N_1\rangle\left[\frac{\langle N_1\rangle}{V} - \frac{1}{\lambda_1^3}\exp\left(\frac{1}{k_BT}\frac{\partial f_l}{\partial l}\right)\right]. \quad l \geq 2 \quad (5.8)$$

In both equations (5.7) and (5.8) the monomer distribution results from the boundary conditions (5.6).

Equation (5.8) is a balance equation. Therefore we find the deterministic kinetics

$$\dot{l} = \alpha l^{2/3}\left[\frac{\langle N_1\rangle}{V} - \frac{1}{\lambda_1^3}\exp\left(\frac{1}{k_BT}\frac{\partial f_l}{\partial l}\right)\right]. \qquad l \geq 2 \qquad (5.9)$$

By introduction of the cluster radius $r_l(t)$ and by a linear expansion of the exponential function in (5.9), results for a single cluster the known kinetic equation /23/:

$$\dot{r}_l(t) = \frac{\alpha}{\lambda_1^3}\frac{d_0}{\left(\frac{4\pi}{3}cN_A\right)^{1/3}}\left[\frac{1}{r_c(t)} - \frac{1}{r_l(t)}\right], \qquad (5.10)$$

where $r_c(t)$ is the time-dependent critical radius

$$r_c(t) = d_0\left(\ln\frac{\langle N_1\rangle}{V}\lambda_1^3 + \frac{A}{kT}\right)^{-1}, \qquad (5.11)$$

d_0 being the capillarity length: $d_0 = 2\sigma\,(c\,N_A k_B T)^{-1}$. The density of the free monomers $\langle N_1\rangle/V$ acts as an effective supersaturation.

(iv) The master equation (4.13) can be reduced to a kinetic equation for single clusters as well. If $P_{N_1+1}(l,t)$ is the probability that a l-mer is present in the environment of N_1 free monomers at time t, we propose the following ansatz:

$$\dot{P}(N_1\ldots N_N,t) = \frac{G_N(t)}{N_2!\ldots N_N!}\prod_{l=2}^{N}\left(P_{N_1+1}(l,t)\right)^{N_l}\;;\quad N_1 = N - \sum_{l=2}^{N} l\,N_l\,. \qquad (5.12)$$

The index N_1+1 of $P_{N_1+1}(l,t)$ labels the number of free monomers at the very beginning of forming the present droplet. It is an invariant during the droplet growth. It yields

$$\dot{P}_{N_1+1}(l,t) = \alpha\Big\{(l-1)^{2/3}\frac{N_1+1}{V}P_{(N_1+1)+(l-1)}(l-1,t)$$

$$+\frac{1}{\lambda_1^3}(l+1)^{2/3}\exp\left(\frac{f_{l+1}-f_l}{k_BT}\right)P_{(N_1-1)+(l+1)}(l+1,t)\Big\}$$

$$-\alpha\, l^{2/3}\Big\{\frac{N_1}{V} + \frac{1}{\lambda_1^3}\exp\left(\frac{f_l-f_{l+1}}{k_BT}\right)\Big\}P_{N_1+1}(l,t)\,, \qquad (5.13)$$

where

$$\dot{G}_N(t) = \alpha_1 G_N(t)\left[2^{2/3}\frac{1}{\lambda_1^3}\exp\left(\frac{f_2}{k_BT}\right)P_{(N_1-2)+2} - \frac{N_1(N_1-1)}{V}\right]. \qquad (5.14)$$

Equation (5.14) is the balance equation for the probability of the free monomers and realizes certain physical assumptions. In the considered case it implies the solution of the equations (5.13) and (5.14) under the constraints (1.3). Another possibility where only one single cluster performs the nucleation process is shown in ref. /17/. If we assume N_1 = const. it follows from (5.13) the classical Becker-Döring theory /5/.

References

1. A.C. Zettlemoyer (Ed.): Nucleation (Dekker, New York 1969); Nucleation Phenomena (Wiley, New York 1977)
2. F.F. Abraham: Homogeneous Nucleation Theory (Academic Press, New York 1974)
3. J.D. Gunton, M. San Miguel, P.S. Sahni, in: C. Domb (Ed.) Phase Transitions and Critical Phenomena (Academic Press, New York 1983)
4. M. Volmer: Kinetik der Phasenbildung (Steinkopf, Dresden 1939)
5. R. Becker, W. Döring: Ann.Phys. 24, 719 (1935)
6. J.L. Frenkel: Kinetic Theory of Liquids (Dover, New York 1955)
7. J.B. Zeldovich: J.Exp.Teor.Phys. UdSSR 12, 525 (1942)
8. J.W. Cahn, J.E. Hilliard: J.Chem.Phys. 28, 258 (1958); 30, 1121 (1959); 31, 688 (1959)
9. J.S. Langer: Kinetics of metastable states, in: Lecture Notes in Physics 132 (Springer, Berlin Heidelberg New York 1980)
10. J.S. Langer, A.J. Schwartz: Phys.Rev. A21, 948 (1980)
11. K. Binder, D. Stauffer: Adv. Phys. 25, 343 (1976)
12. K. Binder: Spinodal decomposition, in: Lecture Notes in Physics 132 (Springer, Berlin Heidelberg New York 1980)
13. A.S. Pataschinsky, B.I. Shumilo: JETP 77, 1417 (1979); FTT 22, 2609 (1980)
14. R. Becker: Theorie der Wärme (Springer, Berlin Heidelberg New York 1961)
15. W. Ebeling, D. Kremp, W.D. Kraeft: Theory of bound states (Akademie-Verlag, Berlin 1976; Isdatelstwo Mir, Moscow 1979 (in russ.))
16. H. Haken: Synergetics - An Introduction (Springer, Berlin Heidelberg New York 1978)
17. W. Ebeling: Sitzungsberichte der AdW der DDR, 22N, 33 (1981)
18. J.L. Lebowitz, D.G. Bergmann: Ann.Phys. (N.Y.) 1, 1 (1957)
19. S. Karlin, M. Taylor: Stochastic Processes (Mc Graw, New York 1974)
20. H.K. Janssen:Z.Phys. 270, 67 (1974)
21. W. Ebeling, R. Feistel: Physik der Selbstorganisation und Evolution (Akademie-Verlag, Berlin 1982)
22. C.W. Gardiner: Handbook of Stochastic Methods (Springer, Berlin Heidelberg New York 1984)
23. R.C. Cesai, S.W. Koch, F.F. Abraham: Physica 118A, 136 (1983)

On the Properties of the Nonequilibrium Potential Near Bifurcation Points

A.V. Tolstopjatenko[1] *and L. Schimansky-Geier*[2]

[1]Department of Physics, MIIGAiK, SU-103 064 Moscow, USSR
[2]Humboldt-Universität Berlin, Sektion Physik, Bereich 04, Invalidenstraße 42, DDR-1040 Berlin, GDR

1. Introduction

There has been an increased interest in the effects of Gaussian noise on dissipative dynamical systems described by the kinetic equations /1-8/

$$\dot{x}_i = f_i(\underline{x}) + \sqrt{2\varepsilon}\ \xi_i(t) \qquad (1.1)$$

with the stochastic force

$$\langle \xi_i(t) \rangle = 0; \quad \langle \xi_i(t)\, \xi_j(t') \rangle = \delta_{ij}\, \delta(t-t'). \qquad (1.2)$$

Most investigations are performed to get at least information by the knowledge of the corresponding stationary probability density $P^0(x, \varepsilon)$. Some general properties of P^0 were recently developed. GRAHAM and TEL /9.10/ in a series of papers connected the smoothness of the nonequilibrium potential

$$\Phi(\underline{x}, \varepsilon) = -\varepsilon \ln P^0(\underline{x}, \varepsilon) \qquad (1.3)$$

in the weak noise limit $\Phi(x, \varepsilon \to 0) = \Phi^0(x)$ with the integrability condition of an associated Hamilton-Jacobi equation. These authors pointed out that the case of smooth Φ is exceptional for dissipative nonlinear systems. Their study of special examples reveals "that approximative potentials may still provide a very accurate description even in cases where exact potential does not exist" /9/.

Thus in references /3,11,12/ by a local solution approach and by assuming the potential noise-strength- dependent $\Phi = \Phi(\underline{x}, \varepsilon)$, noise-induced transitions for several two-dimensional systems were calculated. The main effect was that in situations where no potential condition holds, even additive Gaussian white noise modifies quite remarkably the stationary probability distribution, contrary to the experience of one-dimensional system where only multiplicative noise exhibits noise-induced transitions /13/.

In the present paper we investigate the ε -dependence of the potential Φ . As usual analytic dependence

$$\Phi(\underline{x}, \varepsilon) = \Phi^0(\underline{x}) + \varepsilon \Phi^{(1)}(\underline{x}) + \ldots = \sum_{n=0}^{\infty} \varepsilon^n \Phi^{(n)}(\underline{x}) \qquad (1.4)$$

is expected. We prove the finiteness of the $\Phi^{(1)}$ for several situations, which is necessary for ansatzes like (1.4) and solution in successive powers of small ε . It will be shown for a special class of bifurcations locally and for a special example globally that $\Phi^{(1)}$ becomes infinitely large. In this case we calculate locally a nonanalytic dependence of Φ on ε /14/.

2. The Small ε-Expansion: Local Results

We consider dynamical systems of the type (1.1) with two degrees of freedom (i=1,2). The stationary probability density $P^0(\underline{x},\varepsilon)$ obeys the stationary Fokker-Planck equation (FPE)

$$-\frac{\partial}{\partial x_\alpha} f_\alpha P^0 + \varepsilon \frac{\partial^2}{\partial x_\alpha \partial x_\beta} D_{\alpha\beta} P^0 = 0 . \qquad (2.1)$$

(We follow here and furtheron the convention that repeated indices have to be summed over.) If the noise intensity ε is small as compared to the scale of the nonlinearity of the dynamical system

$$\varepsilon^{1/2} \ll (\Delta x)^{nl} \sim \min\left\{\frac{f'(x)}{f''(x)} , \frac{f'(x)}{\sqrt{f'''(x)}}\right\} ,$$

the most general method of the FPE is the ε-expansion method with the ansatz (1.4) /15/.

Let us discuss this method in more detail and prove its validity in the vicinity of bifurcation points. For simplicity we assume linear-cubic functions $f_\alpha(\underline{x})$ with a fixed point at $\underline{x}=0$:

$$f_\alpha(\underline{x}) = f_{\alpha\gamma} x_\gamma + f_{\alpha\beta\gamma\delta} x_\beta x_\gamma x_\delta .$$

We are interested in the local square approximation of the potential

$$\Phi^{(n)} = \frac{1}{2} \Phi_{\alpha\beta}^{(n)} x_\alpha x_\beta + O(\underline{x}^3,\ldots); \; n=0,1\ldots$$

and find after successive solution in powers of ε

$$\Phi_{\alpha\beta}^{(0)} = \frac{\sigma_{\alpha\mu}}{D^2+a^2} [\tilde{f}_{\mu\beta} + a f_{\mu\beta}] , \qquad (2.2)$$

$$\Phi_{\alpha\beta}^{(1)} = \frac{\sigma_{\alpha\mu}}{D^2+a^2} [(D_{\mu\nu} - a\sigma_{\mu\nu}) b_{\nu\beta} - c g_{\mu\beta}] , \qquad (2.3)$$

where

$$\sigma_{\alpha\mu} = \begin{pmatrix} 0,-1 \\ 1,0 \end{pmatrix} \qquad D = \det|D_{\alpha\beta}|$$

$$\tilde{f}_{\alpha\beta} = D_{\alpha\mu}\sigma_{\mu\nu} f_{\nu\beta} \qquad a = -\mathrm{tr}|\tilde{f}_{\alpha\beta}|/\mathrm{tr}|f_{\alpha\beta}|$$

$$g_{\alpha\beta} = \frac{2a}{D^2+a^2} \tilde{f}_{\alpha\beta} + \frac{a^2-D^2}{a^2+D^2} f_{\alpha\beta}$$

$$b_{\alpha\beta} = -\frac{1}{2} (h_{\alpha\varrho} + b\sigma_{\alpha\varrho}) \frac{\sigma_{\varrho\mu}(g_{\nu\mu} - 1/2 f_{\varrho\varrho} \delta_{\mu\nu})\sigma_{\nu\beta}}{\det|f_{\alpha\beta}| + 3/4(\mathrm{tr}|f_{\alpha\beta}|)^2}$$

$$b = \frac{1}{2} \frac{\mathrm{tr}|g_{\alpha\mu}\sigma_{\mu\nu} h_{\nu\beta}|}{\mathrm{tr}|f_{\alpha\beta}|} \qquad c = -\frac{\mathrm{tr}|D_{\alpha\mu} b_{\mu\beta}|}{\mathrm{tr}|f_{\alpha\beta}|}$$

The tensor

$$h_{\alpha\beta} = \frac{\partial^2}{\partial x_\alpha \partial x_\beta} \left\{ \frac{\partial}{\partial x_\gamma} (D_{\gamma\mu}\sigma_{\mu\nu} + a\delta_{\gamma\nu}) f_\nu(\underline{x}) \right\} \Bigg|_{\underline{x}=0} \qquad (2.4)$$

vanishes in this local approach identically for linear systems and if the potential condition holds /15,16/. In these case $\Phi^{(1)}$ vanishes as well. Let us investigate the value of $\Phi^{(1)}$. Obviously it remains finite as long as $\mathrm{tr}|f_{\alpha\beta}| = \lambda_1 + \lambda_2$ does not vanish, where the λ_i are the eigenvalues of the linearized systems. It is the region far from bifurcation points. For the vicinity of bifurcation points we find

1) In the case that both eigenvalues are real and only one vanishes, $\Phi^{(1)}$ is finite valued. Noise-induced transitions occur for noise intensities larger than the critical $\varepsilon_c \sim \lambda_i \to 0$.
2) If the λ_i are complex conjugated and only the real part vanishes, (Hopf-bifurcation) the limit to the bifurcation point of $\Phi^{(1)}$ remains finite. Noise-induced transitions are found for the same range as in 1).
3) If both eigenvalues vanish mutually $\lambda_1 \sim \lambda_2 \sim \lambda \to 0$ and $\varepsilon > \lambda^2$, the $\Phi^{(1)}$ diverges as $1/\lambda$.

It will be more divergent if one takes into consideration the quadratic nonlinear terms $f_\alpha(\underline{x})$ ($\Phi^{(1)}_{\alpha\beta} \sim 1/\lambda^2$ if $\varepsilon > \lambda^3$) and a dependence of the diffusion matrix on $\underline{x}$. If in saddle points $\lambda_1 = -\lambda_2$ and $\det|f_{\alpha\beta}|=0$, the $\Phi^{(1)}_{\alpha\beta}$ will diverge as well.

3. The Method of Reducing Nonpotential Systems to Potential Systems

Let us consider a second approach of solving the stationary FPE- the method of reducing the FPE to the potential form /11,12/. In this approach we will not make use of the ansatz (1.4).

The FPE is written down in the form of a continuity equation

$$\frac{\partial P}{\partial t} - \mathrm{div}\, J = 0, \text{ or } \frac{\partial \Phi}{\partial t} = v_\alpha \frac{\partial \Phi}{\partial v_\alpha} - \varepsilon \frac{\partial v_\alpha}{\partial x_\alpha} = 0 \;, \tag{3.1}$$

where J is the probability flux with the velocity v

$$J_\alpha = v_\alpha P, \quad v_\alpha(x) = K_\alpha(x) + D_{\alpha\beta}(x)\frac{\partial \Phi}{\partial x_\beta} - \varepsilon \frac{\partial}{\partial x_\beta} D_{\alpha\beta}(x) \;. \tag{3.2}$$

In this chapter we suppose multiplicative noise $\sim \sqrt{2\varepsilon}\, g^i_\alpha(x)\, \xi_i(t)$ and interpret the Langevin equations in the Stratonovich sense. Therefore it holds

$$K_\alpha = f(x) + \varepsilon \frac{\partial g^i_\alpha(x)}{\partial x_\beta} g^i_\beta(x); \; D_{\alpha\beta} = g^i_\alpha(x) g^i_\beta(x) \;. \tag{3.3}$$

Further we use that the addition of an antisymmetric matrix $d_{\alpha\beta}(\underline{x}) = -d_{\beta\alpha}(\underline{x})$ to the diffusion matrix $D_{\alpha\beta}(\underline{x})$ does not change the FPE and the probability density but changes the probability flux J and their velocity v /5,17/. We define the matrix $d_{\alpha\beta}(\underline{x})$ so that the J_α and v_α vanish

$$\begin{aligned} D_{\alpha\beta} &\to \tilde{D}_{\alpha\beta} = D_{\alpha\beta} + d_{\alpha\beta} \;, \\ P &\to \tilde{P} \equiv P \;, \; J_\alpha \to \tilde{J}_\alpha = 0, \; v_\alpha \to \tilde{v}_\alpha = 0. \end{aligned} \tag{3.4}$$

Afterwards the transformed system is regarded to obey a potential condition. It is the equation of definition for the matrix elements $d_{\alpha\beta}(\underline{x})$. The nonequilibrium potential $\Phi(\underline{x}, \varepsilon)$ follows from integrating over an arbitrary path

$$\Phi(\underline{x}, \varepsilon) = - \int dx_\alpha \, \tilde{D}^{-1}_{\alpha\beta}(\underline{x})(K_\beta(\underline{x}) - \varepsilon \frac{\partial \tilde{D}_{\beta\gamma}(\underline{x})}{\partial x_\gamma}) \tag{3.5}$$

if

$$\frac{\partial}{\partial x} \tilde{D}^{-1}_{\gamma\beta}(K_\beta - \varepsilon \frac{\partial \tilde{D}_{\beta\gamma}}{\partial x_\gamma}) = \frac{\partial}{\partial x_\gamma} \tilde{D}^{-1}_{\mu\beta}(K_\beta - \varepsilon \frac{\partial \tilde{D}^{-1}_{\beta\gamma}}{\partial x_\gamma}) . \tag{3.6}$$

Equation (3.6) is the generalized potential condition for determining the unknown $d_{\alpha\beta}$.

The occurrence of the antisymmetric diffusion part can be explained also by integrating (3.1) with regard to vanishing probability flux at infinity. It yields the special solution

$$v_\alpha(\underline{x}) = -d_{\alpha\beta}(x) \frac{\partial \Phi}{\partial x_\beta} + \varepsilon \frac{\partial}{\partial x_\beta} d_{\alpha\beta}(x); \quad d_{\alpha\beta} = -d_{\beta\alpha} \tag{3.7}$$

and subsequently

$$\tilde{v}_\alpha = v_\alpha + d_{\alpha\beta}(x) \frac{\partial \Phi}{\partial x_\beta} + \varepsilon \frac{\partial}{\partial x_\beta} d_{\alpha\beta}(x) = 0. \tag{3.8}$$

We show now how the method of reducing nonpotential systems to the potential case can be used for the determination of stochastic bifurcations. Returning back to the Langevin equations

$$\dot{x}_\alpha = f_\alpha(x) + \sqrt{2\varepsilon} \; g^i_\alpha(x) \, \xi_i(t) , \tag{3.9}$$

we consider two situations:

1. The potential case $J_\alpha(x) \equiv 0$:

In this case we obtain explicitly

$$\frac{\partial \Phi}{\partial x_\alpha} = D^{-1}_{\alpha\beta}(x)(f_\beta(x) - \varepsilon g^i_\alpha(x) \frac{\partial g^i_\beta(x)}{\partial x_\beta}) . \tag{3.10}$$

If the inverse diffusion matrix $D^{-1}_{\alpha\beta}$ exists and has no singularities, the localization and the stability properties of the extremal points are identical to the stationary points of the auxiliary dynamical system /19/

$$\dot{x}_\alpha = \bar{f}_\alpha(x) = f_\alpha(x) - \varepsilon g^i_\alpha(x) \frac{\partial g^i_\beta(x)}{\partial x_\beta} . \tag{3.11}$$

The shift between extremal ($\bar{f}_\alpha(x^{ex})=0$) and stationary points ($f_\alpha(x^0)=0$) in potential systems is due to multiplicative noise. It is described by simple renormalization of deterministic equations.

2. The nonpotential case $J_\alpha(x) \neq 0$

One can prove that in this case the influence of external noise upon the localization and stability properties of extremal points is described again with the help of renormalized dynamical systems

$$\dot{x}_\alpha = \tilde{f}_\alpha(x) = f_\alpha(x) - \varepsilon g^i_\alpha(x) \frac{\partial g^i_\beta(x)}{\partial x_\beta} - \varepsilon \frac{\partial d_{\alpha\beta}(x)}{\partial x_\beta} . \tag{3.12}$$

We see that the effect of the noise on the original system is twofold in nonpotential systems: Displacements and changes of stability are caused by multiplicative noises as well as by the non-zero probability flux which results in the existence of the $d_{\alpha\beta}(x)$.

We have obtained the striking effect that noise-induced transitions may even be found in nonpotential systems with additive noises.

The main problem in this approach is the calculation of the anti-symmetrix matrix $d_{\alpha\beta}(x)$. For instance, for two-dimensional systems $d_{\alpha\beta}(x)$ may be represented by

$$d_{\alpha\beta}(x) = \chi(x)\sigma_{\alpha\beta} \; ; \quad \sigma_{\alpha\beta} = \begin{pmatrix} 0 & -1 \\ 1 & 0 \end{pmatrix} , \tag{3.13}$$

where $\sigma_{\alpha\beta}$ is the unit antisymmetric tensor.

In this case the problem of solving the potential condition reduces to the search for one scalar function $\chi(x)$ from the equation

$$\frac{\partial}{\partial x}\left(\frac{D_{\alpha\varrho}(x)\sigma_{\varrho\beta}+\chi(x)\,\delta_{\alpha\beta}}{\det|D_{\alpha\beta}(x)|+\chi^2(x)}\right)\left(K_\beta(x)-\varepsilon\frac{\partial D_{\beta\gamma}(x)}{\partial x_\gamma}-\varepsilon\,\sigma_{\beta\gamma}\frac{\partial\chi(x)}{\partial x}\right)=0 . \tag{3.14}$$

The potential $\Phi(x)$ is given by

$$\Phi(x)=\int dx\,\frac{\sigma_{\gamma\mu}D_{\mu\nu}(x)\sigma_{\nu\beta}+\chi(x)\sigma_{\alpha\beta}}{\det|D_{\alpha\beta}(x)|+\chi^2(x)}\left(K_\beta(x)-\varepsilon\frac{\partial D_{\beta\gamma}(x)}{\partial x_\gamma}-\varepsilon\sigma_{\beta\gamma}\frac{\partial\chi(x)}{\partial x_\gamma}\right) . \tag{3.15}$$

4. A Local Approach to the Solution of Two-Dimensional Fokker-Planck Equations

As was mentioned above, the ε-expansion method breaks down in the vicinity of certain bifurcation points. So one should look for a method which is not connected with an assumption of the dependence on the noise intensity ε . We want to propose in this report one such method - the local approximation by corresponding potential systems.

Looking for local solutions in the neighbourhood of a fixed point $\underline{x}^0$ (for example - a stationary or an extremal one) we expand the antisymmetric matrix $d_{\alpha\beta}(x)$ or the functions $\chi(x)$ in a series of polynomials around the fixed point $\underline{x}^0$

$$\chi(x) = \chi + \chi_\alpha x_\alpha + \frac{1}{2}\chi_{\alpha\beta}x_\alpha x_\beta + \dots . \tag{4.1}$$

Then we obtain from (3.14) an algebraic system for the unknown coefficients of $\chi(x)$. This system is certainly an unclosed one, unless the noise intensity is small enough.

Truncating the series (4.1) we can make the algebraic system closed and find all the coefficients of function $\chi(x)$. The actual ε -dependence of $\chi(x)$ and therefore of $\Phi(x)$ follows from the considered dynamics of the system.

The physical meaning of the truncation procedure consists in the replacing of the initial system with one locally closed to its potential.

Denoting by N the number of powers retained in (4.1), we can find the connection between the kinetic coefficients of the initial system $K_\alpha(x)$ and the transformed local one $\tilde{K}^N_\alpha(x)$

$$\tilde{K}^N_\alpha(x) = -(D_{\alpha\beta}+\chi^N(x)\sigma_{\alpha\beta})\frac{\partial\Phi^N}{\partial x_\beta}+\varepsilon\frac{\partial(D_{\alpha\beta}+\chi^N(x)\sigma_{\alpha\beta})}{\partial x_\beta} ,$$

$$|K_\alpha(x) - \tilde{K}^N_\alpha(x)| = 0\left[\max(x^{N+2}, \varepsilon x^N)\right] . \tag{4.2}$$

One has to choose this number N which makes the kinetic coefficients of the initial and the transformed system topologically equivalent in the neighbourhood of the investigated point neglecting noisy effects.

5. Examples

As a first example where one eigenvalue vanishes, we consider a reaction model under the influence of an Ornstein-Uhlenbeck process /13/. Written as a Markovian system it reads

$$\dot{x} = -\alpha x + (x^2 - \tfrac{1}{4})y$$

$$\dot{y} = -\gamma y + \sqrt{2\varepsilon}\ \xi(t); \ \langle y(t)y(t')\rangle = \frac{2\varepsilon}{\gamma}\, e^{-\gamma|t-t'|} \quad . \tag{5.1}$$

Here $\alpha = \tau_{rel}^{-1}$ is the reciprocal relaxation time of the chemical reaction. y(t) is the Ornstein-Uhlenbeck process with reciprocal correlation time $\gamma = \tau_c^{-1}$ and we assume $\gamma \gg \alpha$. From the local analysis we find the stationary probability density $\Phi^{(0)} + \varepsilon\Phi^{(1)}$ around the single stationary point of the deterministic system $x^0 = y^0 = 0$. It possesses the properties: If

$$\varepsilon < \varepsilon_c = \frac{2\alpha\gamma(\alpha+\gamma)^2(\alpha+3\gamma)(\gamma+3\alpha)}{17\alpha^3+38\alpha^2\gamma+22\alpha\gamma^2+3\gamma^3} , \tag{5.2}$$

there exists a single maximum above the deterministic stationary state $x^0 = y^0 = 0$. If $\varepsilon > \varepsilon_c$ this point becomes unstable and two maxima appear

$$x_1^{ex} \approx \pm \frac{1}{2}\sqrt{\frac{\varepsilon-\varepsilon_c}{\varepsilon_c}} \quad ; \quad \left(\frac{\varepsilon-\varepsilon_c}{\varepsilon_c} \ll 1\right) . \tag{5.3}$$

Comparing with the critical white-noise-density $D_w^c = 2\alpha$, we find

$$D_{ou}^c = \frac{\varepsilon_c}{\gamma^2} \approx D_w^c\ (1 - 2\,\frac{\alpha}{\gamma} + O((\frac{\alpha}{\gamma})^2)) \ . \tag{5.4}$$

Secondly we consider the Van der Pol oscillator

$$\dot{x} = y, \ \dot{y} = -x + (\alpha - y^2)y + \sqrt{2\varepsilon}\ \xi(t) \ . \tag{5.5}$$

At $\alpha = 0$ a Hopf bifurcation takes places and a stable limit cycle for $\alpha > 0$ appears. Though a nonanalytic dependence of χ on ε in agreement with the results section 2 the ε-expansion method is valid closely to the bifurcation point /11/. The stationary probability density is given by

$$P^0(x,y) \sim \exp\left[\frac{1}{\varepsilon}(\frac{\alpha}{2}(x^2+y^2) + \frac{3}{4}\,\varepsilon xy - \frac{3}{16}(x^2+y^2)^2)\right] \ . \tag{5.6}$$

In the scaling region $\alpha \sim \varepsilon^{1/2}$, $x,y \sim \varepsilon^{1/4}$ this result agrees with that obtained from calculations using slowly changing amplitudes /4,5/. For smaller values $\alpha \sim \varepsilon$ the probability density becomes strongly asymmetric, due to the second term in the exponent. For $|\alpha| \leq 3/8\,\varepsilon$, this asymmetry results in the appearance of a bimodal-like distribution with two maxima above $x^0 = \pm\sqrt{1/3(4\alpha - 3/2\,\varepsilon)}$, $y^0 = -x^0$. The scenario of a stochastic Hopf bifurcating system differs quite strongly from the deterministic case, even in the presence of additive noise only /3,8,12/.

As the third example we investigate the model

$$\dot{x} = -\gamma x - xy + \sqrt{2\varepsilon}\, \xi_1(t); \quad \dot{y} = -\gamma y + x^2 + \sqrt{2\varepsilon}\, \xi_2(t) . \tag{5.7}$$

Obviously it is the case 3) of section 2. If $\gamma \to 0$ the stable attractor $x^0 = y^0 = 0$ degenerates in an unstable manifold $x=0$; $y<0$, and in a stable one $x=0$; $y>0$. We introduce (5.7) because it allows a global solution of the FPE to first order in ε for $\gamma > 0$

$$\Phi(\underline{x}, \varepsilon) = \frac{\gamma}{2} r^2 + \varepsilon \ln(\cosh \frac{r}{\gamma} - \sinh \frac{r}{\gamma} \sin \varphi) + O(\varepsilon^2) . \tag{5.8}$$

$$(r = \sqrt{x^2+y^2} \; , \quad \varphi = \arctan(y/x))$$

In agreement with the local analysis of section 2 we find that the second term becomes dominant if γ is small. For arbitrary $r/\gamma \gg 1$ it yields

$$\Phi \approx \frac{\gamma}{2} r^2 + \varepsilon \frac{r}{\gamma} \; , \quad \varphi \neq \frac{\pi}{2}$$

$$\Phi \approx \frac{\gamma}{2} r^2 - \varepsilon \frac{r}{\gamma} \; , \quad \varphi = \frac{\pi}{2} . \tag{5.9}$$

The second term is important for

$$\varepsilon > \gamma^2 r \gg \gamma^3 > 0 . \tag{5.10}$$

From the local approach we obtain a strong dependence of the shape of the probability distribution on the parameter $\delta = \varepsilon^{1/3}/\gamma$. The expansion coefficients (4.1) of Ψ are found to be

$$\Psi_0 = \Psi_2 = \Psi_{11} = \Psi_{22} = 0 \tag{5.11}$$

and Ψ_1, Ψ_{12} obey the equations

$$2\varepsilon \Psi_1^3 + 3\gamma \Psi_1 + 3 = 0 ,$$

$$\Psi_{12} = \frac{2\Psi_1^4}{3} \frac{\varepsilon}{2\gamma + 3\Psi_1^2 \varepsilon} . \tag{5.12}$$

For algebraic simplicity we have expanded instead of φ its inverse function $\Psi = \varphi^{-1}$. The Ψ_0 , Ψ_α , $\Psi_{\alpha\beta}$ are the resulting coefficients of Ψ.

Two regions depending on the value of δ should be distinguished:

a) extensive region ($\delta \ll 1$): $\Psi_1 \approx -\frac{1}{\gamma}$; $\Psi_{12} \approx 0$

In agreement with the former result (5.8) it follows from (3.17)

$$\Phi(x) \approx \frac{\gamma}{2} r^2 - \varepsilon\left(\frac{x_2}{\gamma} - \frac{x_1^2}{2\gamma^2}\right) + \text{nonlinear terms}; \; (\frac{r}{\gamma} \ll 1) . \tag{5.13}$$

It is the region where the ε-expansion is valid.

b) fluctuation region ($\delta \gtrsim 1$): $\Psi_1 \approx -\sqrt[3]{\frac{3}{2\varepsilon}}$; $\Psi_{12} \approx \frac{1}{3}\sqrt[3]{\frac{2}{3\varepsilon^2}}\,(1+O(\delta))$

The application of the local transformation method results here in a nonanalytic dependence of the nonequilibrium on the noise strength /18/

$$\Phi(x) \approx -\left(\frac{3\varepsilon^2}{2}\right)^{1/3} y + \left(\frac{2\varepsilon}{3}\right)^{1/3}\left(\frac{7}{6}x^2 + \frac{1}{3}y^2\right) + O(x^3, \delta, \gamma) . \tag{5.14}$$

The scaling of this expression $\Phi(x) = \varepsilon\Phi(\frac{x}{\varepsilon^{1/3}}, \delta)$, $\Psi = \Psi(x/\varepsilon^{1/3}, \delta)$

is inherent in the initial FPE.
Writing $\tilde{x}_{1,2} = x_{1,2}/\varepsilon^{1/3}$ we get

$$\frac{\gamma}{\varepsilon^{1/3}}\frac{\partial}{\partial\tilde{x}_\alpha}\tilde{x}_\alpha P + \frac{\partial}{\partial\tilde{x}_1}\tilde{x}_1\tilde{x}_2 P - \frac{\partial}{\partial\tilde{x}_2}\tilde{x}_1^2 P + \frac{\partial^2 P}{\partial\tilde{x}_1^2} + \frac{\partial^2 P}{\partial\tilde{x}_2^2} = 0. \qquad (5.15)$$

We investigate the properties of the renormalized dynamical system (3.14). It yields

$$\dot{x} = -\gamma x - xy + \varepsilon\frac{\partial\Psi}{\partial y}\ ;\ \dot{y} = -\gamma y + x^2 - \varepsilon\frac{\partial\Psi}{\partial x}. \qquad (5.16)$$

Transforming linearly in the stationary states of (5.16)
$x^0=0$; $y^0=-\varepsilon\Psi_1/(\gamma+\varepsilon\Psi_{12})$ we find ($\tilde{x}=x$; $\tilde{y}=y-y^0$)

$$\dot{\tilde{x}} = -\tilde{\gamma}_1\tilde{x} - \tilde{x}\tilde{y};\ \dot{\tilde{y}} = -\tilde{\gamma}_2\tilde{y} + \tilde{x}^2 \quad , \qquad (5.17)$$

with

$$\tilde{\gamma}_1 = \gamma - \varepsilon\Psi_{12} + y^0,\quad \tilde{\gamma}_2 = \gamma + \varepsilon\Psi_{12}\ , \qquad (5.18)$$

which are both positive as long as $\gamma > 0$. Therefore we find in both regions shift of the maximum along the y-axis.

6. Conclusions

We have discussed the behaviour of the nonequilibrium potential near several bifurcation points. At common bifurcations like Hopf bifurcation or the generation of a singularly stationary state, the effect of the noise results if $h_{\alpha\beta} \neq 0$ in the appearance of a finite contribution in Φ . It depends linearly on the noise strength ε and therefore modifications of the bifurcation scenario occur if the bifurcation parameter $\chi \sim \varepsilon$.

This picture changes drastically in the vicinity of complicated structurally unstable situations where Tr $f_{\alpha\beta} = 0$ and $\det|f_{\alpha\beta}| = 0$. In most of these cases the stationary probability density due to the existence of extended stationary manifolds at the bifurcation point does not exist. In our analysis this is reflected by the dominating increase of the $\Phi^{(1)}$ in an assumed analytical series (1.4): $\Phi = \Phi^{(0)} + \varepsilon\Phi^{(1)} + \ldots$ This increase, as was shown in section 5, is not caused by the local solution approach. By the second local solution approach, nonanalytic dependence of the potential on the noise intensity was observed.

The situation which happens near this complicated bifurcation remains quite unsatisfactory. On the one side an assumed perturbation series (1.4) breaks down due to the large increase of the second approximation. It makes the considerations of only $\Phi^{(0)}$ near this bifurcation point at least problematic. On the other hand, the results of the local approach near global bifurcations obviously are far from being sufficient, and may be used only as a hint for further research. The nonanalytic dependence resulting here from the interaction of noise and nonlinear terms in the presence of low linear modes must be reestablished, possibly by computer experiments or by functional methods going beyond a saddle point, or a Gaussian approximation.

References

1. H. Haken: Synergetics - An Introduction, 2nd ed., Springer Ser. in Synergetics Vol. 1 (Springer, Berlin Heidelberg New York 1978)
2. H. Risken: The Fokker-Planck-Equation, Springer Ser. in Synergetics Vol. 18 (Springer, Berlin Heidelberg New York 1984)
3. H. Malchow, L. Schimansky-Geier: Noise and Diffusion in Bistable Nonequilibrium Systems, Teubner-Texte zur Physik, Bd. 4, (Teubner, Leipzig 1985)
4. W. Ebeling, Yu.L. Klimontovich: Selforganization and Turbulence in Liquids, Teubner-Texte zur Physik, Bd. 2 (Teubner, Leipzig 1984)
5. W. Ebeling, H. Engel-Herbert: Physica 104A, 378 (1980); Adv. Mechanics 5, 41 (1982)
6. F. Baras, M. Malek-Mansour, C. Van den Broeck: J.Stat.Phys. 28, 577 (1982)
7. K.A. Wiesenfeld, E. Knobloch: Phys.Rev. A26, 2946 (1982)
8. K.H. Hoffmann: Z.Phys. B49, 245 (1982)
9. R. Graham, T. Tel: Phys.Rev.Lett. 52, 9 (1984)
10. R. Graham, T. Tel: J.Stat.Phys. 35, 729 (1984); 37, 709 (1984); Phys.Rev. A31, 1109 (1985)
11. L. Schimansky-Geier, A.V. Tolstopjatenko, W. Ebeling: Phys. Lett. 108A, 329 (1985)
12. A.V. Tolstopjatenko, L. Schimansky-Geier: Acta Phys.Pol. A66, 197 (1984); W. Ebeling, H. Herzel, W. Richert, L. Schimansky-Geier: ZAMM in print; W. Ebeling, L. Schimansky-Geier: Fluid Dyn. Transactions 12, 7 (1985)
13. W. Horsthemke, R. Lefever: Noise-Induced Transitions-Theory and Applications in Physics, Chemistry and Biology, Springer-Ser. in Synergetics Vol. 15 (Springer, Berlin Heidelberg New York 1984)
14. A.V. Tolstopjatenko, L. Schimansky-Geier: Phys.Lett. A submitted
15. R. Graham: in Springer Tracts in Modern Phys., Vol. 66, p. 1 (Springer, Berlin Heidelberg New York 1973)
16. D. Ryter: J.Phys. A 18, 1111 (1985)
17. M. San Miguel, S. Chaturvedi: Z.Phys. B40, 167 (1980)
18. A.V. Tolstopjatenko: Thesis, Moscow State University, Moscow 1982
19. W. Ebeling: Phys.Lett. 68A, 430 (1978)

Nonlinear Fluctuation-Dissipation Models of Charge Transfer. Nonequilibrium Electrical Noise and 1/f-Noise

G.N. Bochkov

Gorki State University, Prospekt Gagarin 23, SU-603 600 Gorki, USSR

1. Generalized fluctuation-dissipation relations (FDR) /1-4/ are the consequence of 1) invariance of laws of microdynamics with respect to time inversion and 2) conservation of microscopic phase volume. FDR associate functions of nonlinear response (to external dynamic or internal thermal perturbations) with correlation functions of different order in an arbitrary nonequilibrium adiabatically closed system. Using FDR implies that a system contains both the "dynamic subsystem" and "thermostat" or surroundings. Giving concrete expression to such division may be used only when passing from exact FDR to stochastic model.

Consider the system with hamiltonian $\hat{H}_0$ having at the moment $t=0$ distribution function $\hat{\rho} = \exp\left\{\frac{F}{T} - \frac{\hat{H}_0}{T} - Z_\alpha Q_\alpha\right\}$. Let it at $t > 0$ be exposed to external dynamic perturbation: $\hat{H}_0 \longrightarrow \hat{H}(t) = \hat{H}_0 - X_\alpha(t)\hat{Q}_\alpha$. The set of FDRs is equivalent to the following relation for probability functional of the system /1-2/:

$$P[Q(\tau);J(\tau);X(\tau);Z]\exp\left\{-\int_0^t \left[Z_\alpha + \frac{X_\alpha(\tau)}{T}\right] J_\alpha(\tau)d\tau\right\}$$
$$= P[\varepsilon Q(t-\tau);\ -\varepsilon J(t-\tau);\ \varepsilon X(t-\tau);\ \varepsilon Z]\,, \tag{1}$$

where $J(\tau) = d/d\tau\, Q(\tau)$; $\varepsilon_\alpha = \pm 1$ depending on the evenness of quantity Q_α with respect to time inversion. According to (1) when $Z_\alpha=0$ (initial state is thermodynamic equilibrium) the ratio of probabilities of mutually time-inversed processes are completely defined by the work of perturbing forces $X_\alpha(t)$.

2. Consider nonequilibrium fluctuations of current $J(t)$ in dissipative electrical two-port network. Let the system "two-port network & surroundings" be equilibrium when $t < 0$ constant voltage $X(t)=X$ being applied at the moment $t=0$. Let us use the following model: 1) fluctuations of current are wide-band (white) noise, statistical characteristics of which are subjected to low-frequency fluctuations; 2) when $t > 0$ stationary state is established (this supposition means disregard of Joule heating). FDR enable us to associate /1,2/ volt-ampere characteristic (average current) $\bar{J}(X)$ ($\bar{J} \equiv \langle J\rangle$) and spectral density of nonequilibrium white noise:

$$\bar{J}(X) = S(X)\int_0^\infty \mathrm{th}\left(\frac{qX}{2T}\right) V(q;X)\,\frac{dq}{q}\,, \tag{2}$$

where $V(q;X) \geq 0$; $\int_0^\infty V(q;X)\,dq=1$. Function $V(q;X)$ is defined by the statistics of charge transfer (by the white noise statistics). In case of Gaussian statistics $V(q;X) = \delta(q)$ and

$$J(X) = \frac{X}{2T}\,S(X), \qquad T_n(X) = T\,\frac{\bar{J}}{X}\left(\frac{d\bar{J}}{dX}\right)^{-1}, \tag{3}$$

where $T_n(X)$ is the effective temperature of the noise. In case of Poissonian statistics $V(q;X) = \delta(q-q_0(X))$. Note the inequality stemming from (2)

$$2T\frac{\bar{J}(X)}{X} \leqslant S(X) \leqslant 2q_0(X)|\bar{J}(X)| \left[1-\exp\left(-\frac{q_0(X)|X|}{T}\right)\right]^{-1},$$

where $q_0(X) \equiv \int_{-\infty}^{\infty} qV(q;X)dq$ is the charge transfered in typical elementary act.

Condition $\frac{q_0(X)|X|}{T} \sim 1$ delimits the regions of weak and highly nonequilibrium states. It means that the energy dissipated in each elementary act is of the order of T. In a weak nonequilibrium state there is formula (3), i.e. Gaussian model of noise may be used. In case of highly nonequilibrium state approximately

$$S(X) = q_0(X)|J(X)| \; ; \quad T_n(X) = \frac{1}{2}q_0(X)|\bar{J}|\left(\frac{d\bar{J}}{dX}\right)^{-1}$$

(when having power law volt-ampere characteristic $T_n(X) \sim q_0(X)|X|$).

3. When analyzing low-frequency fluctuations and 1/f-noise in a weak nonequilibrium state the so-called cubic, or four-indices, FDR /1,3,4/ are of some help. From (1) there follows formula

$$\langle J(t)\rangle_X = \frac{X}{T}\int_0^t \langle J(t),J(t')\rangle_X dt' - \frac{X^3}{6T^3}\iiint_0^t \langle J(t),J(t_1),J(t_2),J(t_3)\rangle_X d^3t + \ldots, \tag{4}$$

in which members $\sim X^n$, $n \geqslant 5$ are omitted. Cancelling cubic, with respect to X, members in (4) we get the relation associating 1) cubic part of the average response, 2) quadratic, with respect to X, part of nonequilibrium correlation function and 3) the fourth-order correlator (cumulant) of equilibrium fluctuations of current which characterizes their nongaussianity and keeps information on power fluctuations of equilibrium white noise. Thus, nonequilibrium low-frequency noise contains two components, one of which is defined by equilibrium fourth cumulant, and the other - by slow nonequilibrium processes - for example, by thermal processes - resulting in nonstationary character of cubic response.

Consider the first component (and 1/f-noise of equilibrium nature) supposing that at $t \gtrsim \tau_\mu$ (where τ_μ is the correlation time of equilibrium noise) stationary state is established. Besides,

$$\langle J(t),J(t')\rangle_X = \langle J(t),J_0(t')\rangle_0 + X^2K_g(t-t') + \ldots ,$$

where $K_g(\tau)$ has the meaning of correlation function of fluctuations of conductivity. From (4) we get the expression for $K_g(\tau)$ in terms of the fourth cumulant

$$K_g(\tau) = \frac{1}{2T^2}\iint_0^\tau \langle J(\tau),J(t'),J(t''),J(0)\rangle_0 \; dt'dt'' . \tag{5}$$

Practically, this formula is a microscopic definition of $K_g(\tau)$ (for $\tau > \tau_\mu$).

4. Let us prove that $K_g(\tau)$ always has a slowly (logarithmically) damping "tail" (caused by natural fluctuations of "intensity" of Brownian motion of charge carriers). Consider current fluctuations in an equilibrium short-circuit two-port network (X=0) and characteristic function

$$\varphi_\tau(iV) = \frac{1}{\tau}\ln\langle\exp[iV\int_{t_0}^{t_0+\tau} J(t')dt']\rangle$$
$$= \int_{-\infty}^{\infty}(\cos vq-1)\frac{S}{q^2}W_\tau(q)dq = -\frac{v^2}{2!}S + \frac{v^4}{4!}B(\tau) - \ldots . \tag{6}$$

Here $t > \tau_\mu$; S is the spectral density of current noise; the magnitude of $B(\tau)$ is expressed simply in terms of the fourth cumulant of current; as the result of (5)

$$\frac{d^2}{d\tau^2} \tau B(\tau) = 24\, T^2 K_g(\tau) ;$$

$\int_{-\infty}^{\infty} W_\tau(q)dq=1$; $\Delta Q(\tau) = \int_{t_0}^{t_0+\tau} J(t')dt'$ is the charge having diffused through two-port network during time τ .

If there are no slow fluctuational phenomena in the system affecting the diffusion of charge, then at $\tau \gg \tau_\mu$ diffusional random process $\Delta Q(\tau)$ should be scale invariant. The character of invariancies clear from low $\langle \Delta Q^2(\tau) \rangle_0 = S\tau$. Therefore diffusion picture does not change when the charge scale Q is extended in λ times while time scale τ being extended in λ^2 times:

$$\lambda^2 \varphi_{\lambda^2\tau}(iv/\lambda) = \varphi_\tau(iv) .$$

However, the invariance should be distorted at characteristic microscales q_0, τ_0. Noncomplex analysis shows $W_\tau(q)$ in (6) to be

$$W_\tau(q) = \frac{(\ln \tau/\tau_0)}{|q| + q_0} F(\frac{q^2}{2S'\tau}) , \tag{7}$$

where $\int_0^\infty F(z)dZ=1$; $F(0)=1$, $q_0^2=S'\tau_0$. From (6), (7) we find, that

$$B(\tau) = 2SS'\tau \,(\ln \tau/\tau_0)^{-1}$$

and when $\tau \gg \tau_0$,

$$K_g(\tau) = \frac{SS'}{6T^2} (\ln \tau/\tau_0)^{-1} . \tag{8}$$

Spectral density $S_{\delta g}(\omega)$ of relative fluctuations of conductivity, when $(\omega\tau_0)^2 \ll 1$, is equal to

$$\bar{g}^{-2} S_g(\omega) \equiv S_{\delta g}(\omega) = \frac{2\pi}{|\omega|} \frac{q_0^2}{3S\tau_0} (\ln|\omega|\,\tau_0)^{-2} . \tag{9}$$

In the case of statistically independent carriers in a homogeneous sample with length L we have

$$q_0 = er_0L^{-1} , \qquad q_0^2(S\tau_0)^{-1} = r_0^2(2D\,\tau_0N)^{-1} ,$$

where N is the number of carriers, D is the diffusion coefficient, r_0 is defined by the length of their free path. Supposing $r_0^2 \simeq 2D\tau_0$, $\tau \sim 10^{-12}$ sec, $\omega/2\pi \sim 1$ c/sec, from (9) we find the estimate of

$$S_{\delta g}(\omega) \simeq \frac{2\pi}{N|\omega|} \cdot 10^{-3}$$

in accordance with empirical Hooge formula $S_{\delta g} \simeq \frac{2\pi\alpha}{N|\omega|}$ where $\alpha = 2\cdot 10^{-3}$ /5/.

Note, that extremely slow damping of correlation (8) does not show the long-term memory of the system about the past. And vice versa, this is the result of the absence of long-term memory, thermodynamic indifference of the system to spontaneous changes of "intensity" of Brownian motion of carriers. Wandering of a single carrier does not have definite coefficient of diffusion at all (the latter characterizes the big set only). The considered 1/f-noise is an integral attribute of real nongaussian Brownian motion, but is not caused by some "slow mechanisms" (in detail see /6/). Statistical characteristics of 1/f-noise are investigated theoretically in /6/ and in particular 1/f-fluctuations of charge carrier mobility. 1/f-noise probability distribution is shown to be different qualitatively from the Gaussian one, the higher the relative noise level, the stronger being nongaussianity quantitatively. Mobility fluctuation being large (i.e. Hooge constant being 10^{-4}), this distribution has a long quadratic "tail" and is very close to asymmetric stable Cauchy distribution. The most probable properties of flicker (1/f) fluctuations are shown to be sufficiently different comparing with the averaged ones.

References

1. G.N. Bochkov, Yu.E. Kuzovlev: Zh.Eksp.Teor.Fiz. 72, 238 (1977) (Sov.Phys.JETP. 45, 125 (1977)) ; Zh.Eksp.Teor.Fiz. 76, 1071 (1979) (Sov.Phys. JETP. 49, 543 (1979))
2. G.N. Bochkov, Yu.E. Kuzovlev: Physica 106A, 443 (1981)
3. G.F. Efremov: Zh.Eksp.Teor.Fiz. 54, 2322 (1968)
4. R.L. Stratonovich: Zh.Eksp.Teor.Fiz. 58, 1612 (1970) (Sov.Phys. JETP. 31, 864 (1970))
5. F.N. Hooge, T.G.M. Kleinpenning, L.K.J. Vandamme: Rep.Progr.Phys. 44, 479 (1981)
6. G.N. Bochkov, Yu.E. Kuzovlev: Usp.Fiz.Nauk 141, 151 (1983) (Sov. Phys.Usp. 26, 829 (1983)) ; Izv.VUZ.Radiofizika 26, 310 (1983), 27, 1157 (1984); Preprint 195; 157 Research Institute of Radiophysics, Gorki 1985

Application of Path Integrals to Stochastic Reaction-Diffusion Equations

A. Förster[1] *and A.S. Michailov*[2]

[1]Humboldt-Universität Berlin, Sektion Physik, Bereich 04, Invalidenstraße 42, DDR-1040 Berlin, GDR
[2]Department of Physics, Moscow State University, SU-117 234 Moscow, USSR

Investigations of reaction-diffusion equations play a central role in the modern theory of selforganization /1/. Such diverse phenomena as dissipative structures and autowave patterns in active media, nucleation effects in bistable systems and biological evolution processes are described by these equations (see, e.g. /2-5/). External noises can essentially alter the dynamical behaviour of nonequilibrium systems, leading to the noise-induced phase transitions /6/. Stochastic reaction-diffusion equations with external random fields are encountered in studies of the explosion instability in fluctuating continuous systems /3,7,8/, in the problem of diffusion on the fluctuating surface /9/, in discussions of stochastic autowave propagation /10,11/ and synchronization phenomena in random systems of selfexcitable oscillators coupled by diffusion /12/, and in stochastic evolutionary models /13,14/.

Today there are some sufficiently powerful theoretical methods which allow the analysis of processes in stochastic reaction-diffusion systems with external noises. These methods include, for example, the diagrammatic perturbation technique /15/. However, they are not suitable in a number of important problems, such as the estimation of the probability for the creation of the droplets of a new phase in continuous bistable media with the distributed external fluctuation parameters, etc. In our opinion, in this and other related problems, where the probability of very rare events has to be estimated, the treatment can be based on the path-integral solution of stochastic reaction-diffusion equations.

The history of application of path-integral methods for solution of stochastic differential equations starts with the classical work by WIENER /16/, which was generalized by ONSAGER and MACHLUP /17/. The solution of the Fokker-Planck equation in terms of the path integrals was first formulated by STRATONOVICH /18/. Extensive studies of path-integral solutions for the general Fokker-Planck equation and stochastic differential equations were performed by GRAHAM (see, e.g. /19,20/), MARTIN, SIGGIA, ROSE /21/ and others (see, e.g. /22, 23,24/). However, the analysis was usually limited to stochastic differential equations with the white noise. Studies of stochastic differential equations with multiplicative colored noises are scarce (see /25,26/).

In the present paper, we start with a short formulation of the path-integral solution for a general reaction-diffusion equation with multiplicative external random fields, characterized by nonvanishing correlation times and correlation lengths. Next, by using this formal solution, we develop an optimal-fluctuation method, suited for estimation of probabilities of very rare events, and illustrate its application by one special example.

1. We consider a general stochastic reaction-diffusion equation

$$\dot{\Psi} = g(\Psi) + h(\Psi)f(\vec{r},t) + D\Delta\Psi , \tag{1}$$

where $f(\vec{r},t)$ is the Gaussian random field,

$$\langle f \rangle = 0, \quad \langle f(\vec{r},t) f(\vec{r}',t') \rangle = \sigma(\vec{r}-\vec{r}', t-t') , \tag{2}$$

while $g(\Psi)$ and $h(\Psi)$ are some functions of the local density $\Psi(\vec{r},t)$. We want to determine the probability measure $P[\Psi(\vec{r},t)]\, D\Psi(\vec{r},t)$ in the functional space of all regular realizations of the time-dependent field $\Psi(\vec{r},t)$. As usual, this is done by starting from the discrete version of our equation and by diminishing, then, the step of discretization. The discretization procedure is not unique. For example, the differential equation

$$\dot{\Psi} = g(\Psi) + h(\Psi) f(t) \tag{3a}$$

takes the discrete form

$$(\Psi_j - \Psi_{j-1})/\varepsilon = g(\tilde{\Psi}_j) + h(\tilde{\Psi}_j) f_j , \tag{3b}$$

where $\tilde{\Psi}_j = (1-\gamma)\Psi_{j-1} + \gamma \Psi_j$ and the parameter γ (such that $0 \leq \gamma \leq 1$) remains arbitrary.

The derivation of the probability distribution was performed by generalization of the methods, followed in refs. /19-26/; the final result is

$$P[\Psi(\vec{r},t)] = \int Dp(\vec{r},t) \exp(\mathcal{H} - \int d\vec{r}\, dt\, p \dot{\Psi}) , \tag{4}$$

$$\mathcal{H} = \frac{1}{2} \int d\vec{r}_1 d\vec{r}_2 dt_1 dt_2\, \sigma(\vec{r}_1-\vec{r}_2, t_1-t_2)(p(1)h(1)-\gamma h'(1)) \cdot (p(2)h(2)-\gamma h'(2)) + \int d\vec{r}_1 dt_1 [-\gamma g'(1) + p(1)(g(1) + D\Delta\Psi)] , \tag{5}$$

where the notations $p(1)=p(\vec{r}_1,t_1)$, $h(1)=h(\Psi(\vec{r}_1,t_1))$, $h'=dh/d\Psi$, etc., are used. The path integral in (4) is understood as a formal limit of multiple integrals. Equation (4) should be supplemented by the corresponding expression for the path differential:

$$\prod_{j,\vec{r}} d\Psi_j(\vec{r}) \longrightarrow D\Psi(\vec{r},t) . \tag{6}$$

Note, that the correct meaning has only the combination $P[\Psi(\vec{r},t)]$ $D\Psi(\vec{r},t)$ that defines the probability measure in the functional space. Therefore, the factorization of the unique measure into the product of P and $D\Psi$ remains sufficiently arbitrary. For example, instead of (4-6), we can choose another alternative and include all the terms that contain the ambiguity factor γ, into the expression for $D\Psi(\vec{r},t)$. When such a choice is taken we have

$$\mathcal{H} = \frac{1}{2} \int d\vec{r}_1 d\vec{r}_2 dt_1 dt_2\, \sigma(\vec{r}_1-\vec{r}_2, t_1-t_2) p(1)p(2)h(1)h(2) + \int d\vec{r}_1 dt_1 p(1)(g(1) + D\Delta\Psi) . \tag{7}$$

It seems natural to expect that the results of calculation of the averaged quantities, such as correlation functions, would not depend on the details of the discretization rule (i.e. on the ambiguity factor γ). The problem was discussed in the literature (see, e.g. /27-29/) for the path-integral solution of the Fokker-Planck equation. At present, there is no consensus. Some authors argue that the dependence on γ ceases in the final expressions for the averaged quantities; others state that only one particular choice of γ leads to the correct expressions.

2. The probability measure $P[\Psi(\vec{r},t)]\, D\Psi(\vec{r},t)$ contains the most

complete information on statistical properties of the time-dependent random field $\Psi(\vec{r},t)$. It allows the evaluation of the correlation functions or the probabilities of transitions among different configurations. However, this is not simple and straightforward; some further approximations should be taken.

The probabilities of extremely rare events can be found by using a version of "the method of an optimal fluctuation", developed /30/ in the theory of the Schrödinger equation with the random stationary potential. When the probability of transition between two certain configurations $\Psi_0(\vec{r})$ and $\Psi_T(\vec{r})$ in a time interval T is extremely small, it can be estimated as a probability of the most probable path connecting these two configurations. Such a path should obey the variational equations

$$\dot{\Psi} = \delta\mathcal{H}/\delta p(\vec{r},t) \ , \qquad \dot{p} = -\delta\mathcal{H}/\delta\Psi(\vec{r},t) \tag{8}$$

that closely resemble the Hamilton equations of classical mechanics. By using the expressions (4), (5) for the stochastic reaction-diffusion equation (1) we obtain from (8)

$$\dot{\Psi} = h\int d\vec{r}_2 dt_2\, \sigma(\vec{r}_1-\vec{r}_2,t_1-t_2)(p(2)h(2)-\gamma h'(2))+g(\Psi)+D\Delta\Psi, \tag{9a}$$

$$-\dot{p} = (ph'-h'')\int d\vec{r}_2 dt_2\, \sigma(\vec{r}_1-\vec{r}_2,t_1-t_2)(p(2)h(2)-\gamma h'(2))-\gamma g''+pg'+D\Delta p\ . \tag{9b}$$

These two partial differential equations should be supplemented by conditions

$$\Psi(\vec{r},t=0) = \Psi_0(\vec{r}) \ , \qquad \Psi(\vec{r},t=T) = \Psi_T(r) \tag{10}$$

and by the boundary conditions. The solution of (9a,b) under the conditions (10) and the boundary conditions gives the most probable path $\Psi_e(\vec{r},t)$ that connects the initial configuration $\Psi_0(\vec{r})$ and the final configuration $\Psi_T(\vec{r})$. When this solution, together with the corresponding expression* for $p_e(\vec{r},t)$, is substituted into (4), we find an estimate for the transition probability

$$P[\Psi_T(\vec{r})/\Psi_0(\vec{r})] = \int_{\Psi_0(\vec{r}),\ t=0}^{\Psi_T(\vec{r}),\ t=T} D\,\Psi(\vec{r},t)P[\Psi(\vec{r},t)]$$

$$\sim\exp\Big[-\frac{1}{2}\int d\vec{r}_1 d\vec{r}_2 dt_1 dt_2\, \sigma(\vec{r}_1-\vec{r}_2,t_1-t_2)p_e(1)h_e(1)p_e(2)h_2(2)$$
$$+\frac{1}{2}\gamma^2\int d\vec{r}_1 d\vec{r}_2 dt_1 dt_2\, \sigma(\vec{r}_1-\vec{r}_2,t_1-t_2)h_e'(1)h_e'(2)-\gamma\int d\vec{r}_1 dt_1 g_e'(1)\Big]. \tag{11}$$

It should be stressed that t=0 is taken here as a genuine initial moment for the time evolution of the non-Markovian process $\Psi(\vec{r},t)$.

The estimate (11) depends on the parameter γ that specifies the particular choice of the discretization rule. Clearly, such dependence is meaningless. Here we again encounter the ambiguity related to the arbitrariness of discretization. In attempt to solve this difficulty, we consider in the next section an example, where the results of our treatment can be compared with the well-known formulas found in the quantum theory of disordered systems.

3. The Schrödinger equation

$$i\hbar\,\partial\Psi/\partial t = -(\hbar^2/2m)\Delta\Psi + U(\vec{r})\Psi \tag{12}$$

* Note, that $p_e(r,t)$ is also uniquely determined as a solution of (9a,b) under the conditions (10) and the boundary conditions.

allows the stationary state solutions of the form $\Psi(\vec{r},\tau) = \Psi_E(\vec{r})\exp(-iE\tau/\hbar)$. When the potential U is random, the values of energy E in the discrete part of the spectrum are also randomly distributed. One of the principal tasks of the quantum theory of disordered systems consists in determination of the probability distribution p(E), if the statistical properties of the random field $U(\vec{r})$ are known. For sufficiently large (negative) values of E, the probability p(E) becomes exponentially small. The expression for p(E) in this region of energy was found by LAX and LIFSHITZ (see /30-31/) for the Gaussian potential $U(\vec{r})$ by using the method of an optimal fluctuation.

It was easily seen that the Schrödinger equation (12) can be rewritten in a form representing a special case of (1). Indeed, if we denote $D = \hbar^2/2m$, $f = -U(\vec{r})$, $t = -i\tau/\hbar$ and perform the analytical continuation to real t, we find

$$\dot{\Psi} = D\Delta\Psi + f\Psi \ . \tag{13}$$

Here the random field f is stationary, but, in order to reach the complete identification with (1), we can treat it as time-dependent with a very large correlation time τ_c, which is much greater than the observation interval T.

To find the probabilities of very rare events, described by (13), we can use the method formulated in the previous section. In this particular case the variational equations (9a,b) for the most probable path are

$$\dot{\Psi} = D\Delta\Psi + \Psi\int d\vec{r}_2\int_0^T dt_2\,\sigma(\vec{r}_1-\vec{r}_2)(p(\vec{r}_2,t_2)\Psi(\vec{r}_2,t_2)-\gamma) \ , \tag{14a}$$

$$-\dot{p} = D\Delta p + p\int d\vec{r}_2\int_0^T dt_2\,\sigma(\vec{r}_1-\vec{r}_2)(p(\vec{r}_2,t_2)\Psi(\vec{r}_2,t_2)-\gamma) \ . \tag{14b}$$

We are interested in determination of the probability of the extremely rare events, such that

$$\Psi(\vec{r},t) = \nu\,\Psi_0(\vec{r})e^{\lambda t} \ , \ \int\Psi_0^2 d\vec{r} = 1 \ , \tag{15}$$

and the parameter λ is very large. To estimate the probability of the exponential growth with very large values of λ, we can take the probability of the optimal (i.e. the most probable) fluctuation that would give rise to such growth.

Note that Ψ_0 is the wavefunction of the bound state with the energy $E = -\lambda$ in the most probable realization of the random potential $U(\vec{r})$ that contains the level of such depth. Hence, this should be the ground state for the optimal fluctuation, and the wavefunction $\Psi(\vec{r})$ does not have zeros, so that we can take it positive everywhere in the volume.

The optimal fluctuation $\Psi(\vec{r},t)$ obeys the variational equations (14a,b). If it has the time dependence (15), the supplementary function $p(\vec{r},t)$ should be taken in the form $p(\vec{r},t) = \nu\,\Psi_0(\vec{r})e^{-\lambda t}$. Then the function $\Psi_0(\vec{r})$ satisfies the equation

$$D\Delta\Psi_0 + T\nu^2\Psi_0\int d\vec{r}_2\,\sigma(\vec{r}_1-\vec{r}_2)(\Psi_0^2(\vec{r}_2)-\gamma/\nu^2) = \lambda\Psi_0 \ , \tag{16}$$

where $\sigma(\vec{r})$ is the pair correlation function for the Gaussian potential $U(\vec{r})$. Suppose that the correlation length l_c of the random field $U(\vec{r})$ is considerably greater than the localization radius r_Ψ which is defined as

$$r_\Psi = \left[\int r^2\Psi_0^2(r)dr\right]^{1/2} \ . \tag{17}$$

Then we can use the approximation

$$\sigma(r) = \sigma + \frac{1}{2}\sigma_2 r^2 + \ldots \tag{18}$$

and (16) reduces to the Schrödinger equation for the harmonical spherical oscillator:

$$-D\Delta\Psi_0 + \frac{1}{2}r^2 T\sigma_2\Psi_0 = (T\nu^2\sigma - \frac{1}{2}T\nu^2\sigma_2 r_\Psi^2 - \lambda - \gamma TC)\Psi_0\ , \tag{19}$$

where $C = \int \sigma(\vec{r})d\vec{r}$.

Since $\Psi_0(\vec{r})$ does not have zeros it should correspond to the ground state of the spherical oscillator. This requirement is essential, because there is a particular relationship that links the "energy" and the localization radius of the ground state with the oscillator frequency. By using this relationship, we find

$$\nu^2 = (\lambda + \gamma TC)/\sigma T\ . \tag{20}$$

Now, when the optimal function $\Psi_0(\vec{r})$ and the parameter ν are known, we can substitute $\Psi = \nu\Psi_0 e^{\lambda t}$ and $p = \nu\Psi_0 e^{-\lambda t}$ into (11). Finally we get

$$p(\lambda) \sim \exp(-(\lambda + \gamma TC)^2/2\sigma)\ . \tag{21}$$

Since $\lambda = -E$, this expression should coincide with the density of levels in the exponentially small "tail" of the spectrum of the Schrödinger equation (12). The formula for the density of levels in this region is well known, it was found independently by many authors and thoroughly verified /30-33/. Equation (21) reduces to such formula, only we take the value of the ambiguity factor $\gamma = 0$.

We conclude that the compatibility with the results of the quantum theory of disordered systems implies a unique form of the probability functional $P[\Psi(\vec{r},t)]$ when the saddle-point approximation is used. This form corresponds to the choice $\gamma = 0$.

4. The method of an optimal fluctuation, proceeding from the variational equations (14), was further applied by us to investigate the properties of the stochastic differential equation (13) with the time-dependent random fluctuation field $f(\vec{r},t)$. This equation describes, as well, the process of stochastic replication and decay, accompanied by diffusion. The probability of extremely rapid exponential bursts of the population of the replicating species characterized by (15) can be estimated for the time-dependent field $f(\vec{r},t)$ by using the same procedure as above. If the exponential growth persists during the time T that is large compared with the correlation time τ_c of the random field $f(\vec{r},t)$, the probability to find such growth with the increment index λ is given by the following expressions:

a) When $l_c \gg r_\Psi$, $\lambda \gg (\tilde{\sigma}/T)^{1/2}$ and $\lambda \gg D/l_c^2$

$$P(\lambda,T) \sim \exp(-T\lambda^2/2\tilde{\sigma})\ , \tag{22}$$

where $\tilde{\sigma} = \int \sigma(0,\tau)d\tau$.

b) When $l_c \ll r_\Psi$, $\lambda \ll D/l_c^2$

$$P(\lambda,T) \sim \exp\left[-(2T\xi_d/\tilde{\sigma}_0)D^{d/2}(\lambda/\varepsilon_d)^{2-d/2}\right.\ . \tag{23}$$

Here $\tilde{\sigma}_0 = \int\sigma(\vec{r},\tau)d\vec{r}d\tau$, d = 1,2,3 is the dimension of the medium:

$\xi_1 = 4/3$, $\xi_2 = 2{,}419$, $\xi_3 = 6{,}182$; $\varepsilon_1 = 1$, $\varepsilon_2 = 0{,}410$, $\varepsilon_3 = 0{,}107$.

References

1. G. Nicolis, I. Prigogine: Self-Organization in Non-Equilibrium Systems (Wiley, New York 1977)
2. W. Ebeling, R. Feistel: Physik der Selbstorganisation und Evolution (Akademie-Verlag, Berlin 1982)
3. L.S. Polak, A.S. Michailov: Self-Organization in Non-Equilibrium Physical-Chemical Systems (Nauka, Moscow 1983) (in Russian)
4. V.I. Krinsky, A.S. Michailov: Autowaves (Znanie, Moscow 1984) (in Russian)
5. Y. Kuramoto: Chemical Oscillations, Waves and Turbulence (Springer, Berlin Heidelberg New York 1985)
6. W. Horsthemke, R. Lefever: Noise-Induced Transitions (Springer, Berlin Heidelberg New York 1984)
7. A.S. Michailov, I.V. Uporov: Uspekhi Fiz.Nauk 144, 79 (1984)
8. A.S. Michailov, I.V. Uporov: Zh.Eksp.Teor.Fiz. 84, 1481 (1983)
9. A.S. Michailov, I.V . Uporov: Physica 112A, 588 (1982)
10. A.S. Michailov, L. Schimansky-Geier, W. Ebeling: Phys.Lett. 96A, 453 (1983)
11. L. Schimansky-Geier, A.S. Michailov, W. Ebeling: Ann.Physik (Leipzig) 40, 277 (1983)
12. A.S. Michailov, A. Engel: Phys.Lett.A, to be published
13. A.S. Michailov: Phys.Lett. A73, 143 (1979); Dokl.Akad.Nauk SSSR 243, 243 (1978)
14. A.S. Michailov: Z.Phys. B41, 277 (1981)
15. A.H. Nayfeh: Perturbation Methods (Wiley, New York 1973)
16. N. Wiener: Journ.Math.and Phys. 2, 131 (1923)
17. L. Onsager, S. Machlup: Phys.Rev. 91, 1505, 1512 (1953)
18. R.L. Stratonovich: Sel.Transl.Math.Stat.Probl. 10, 273 (1971) (translation of the Russian paper in: Proceedings of the 4th Conference on Probability Theory and Mathematical Statistics, Vilnius 1962, p. 471)
19. R. Graham: Springer Tracts in Modern Physics 66, 1 (1973)
20. R. Graham: Lecture Notes in Physics 84, 82 (1978)
21. P.S. Martin, E.D. Siggia, H.A. Rose: Phys.Rev. A8, 423 (1973)
22. U. Dekker, F. Haake: Phys.Rev. A11, 2043 (1975); A12, 1629 (1975)
23. C. DeDominicis, L. Peliti: Phys.Rev. B18, 353 (1978)
24. H.-J. Janssen: Z .Phys. B23, 377 (1976)
25. R. Phythian: J.Phys. A10, 777 (1977)
26. L. Pesquera, M.A. Rodrigues, E. Santos: Phys.Lett. A94, 287 (1983)
27. R. Graham: Z.Phys. B26, 281 (1977)
28. F. Langouche, D. Roekaerts, E. Terapegui: Phys.Rev. D20, 419 (1979)
29. C. Wissel: Z.Phys. B35, 185 (1979)
    H. Haken (Ed.): Chaos and Order in Nature (Springer, Berlin Heidelberg New York 1981)
30. I.M. Lifshitz, S.A. Gredescul, L.A. Pastur: Introduction to the Theory of Disordered Systems (Nauka, Moscow 1982) (in Russian)
31. B. Halperin, M. Lax: Phys.Rev. 148, 722 (1966)
32. L.V. Keldysh, G.P. Proshko: Fiz.Tverd.Tela 5, 3379 (1963)
33. E.O. Kane: Phys.Rev. 131, 79 (1963)

The Method of Green's Functions in Nonequilibrium Statistical Mechanics

D. Kremp, M. Schlanges, and Th. Bornath

Wilhelm-Pieck-Universität Rostock, Sektion Physik, Universitätsplatz 3, DDR-2500 Rostock, GDR

It is well known that Green's functions are powerful quantities for the description of a many-particle system. In this lecture we are going to deal with the description of nonequilibrium many-particle quantum system with the help of real-time Green's functions (GF). The properties of such systems can be described by the one-particle GF

$$g_1(1,1') = \frac{1}{i}\,\mathrm{Tr}\{\varrho\, T(\Psi(1)\,\Psi^+(1'))\}$$
$$= \theta(t_1-t_1')g^>(1,1')+\theta(t_1'-t_1)g^<(1,1') \;, \qquad (1)$$

with ϱ being the (unknown) nonequilibrium density operator and the operator T representing the Wick's time-ordering operation, "1" is the abbreviation for $\vec{r}_1, t_1$. The correlation functions are

$$g_1^<(11') = \pm\frac{1}{i}\langle\Psi^+(1')\,\Psi(1)\rangle \;;\; g_1^>(11') = \frac{1}{i}\langle\Psi(1)\,\Psi^+(1')\rangle \;.$$

In the case of quantum systems it is convenient to use the Wigner distribution function. The connection with the one-particle GF is

$$f(\vec{p},\vec{R},T) = \pm\, i\int d^3r\,\exp(i\vec{p}\vec{r})g^<(\vec{R}+\tfrac{\vec{r}}{2},T;\vec{R}-\tfrac{\vec{r}}{2},T) \;. \qquad (2)$$

In order to determine the GF we start from the Martin-Schwinger hierarchy /1/ of equations of motion for the real-time GF. The first equations are

$$(i\frac{\partial}{\partial t_1} + \frac{\nabla_1^2}{2m})g_1(1,1') = \delta(1-1')\pm i\int d2\; V(1-2)g_2(12,1'2^+) \qquad (3a)$$

$$(i\frac{\partial}{\partial t_1} + \frac{\nabla_1^2}{2m})g_2(12,1'2') = \delta(1-1')g_1(2,2')\pm (1-2')g_1(2,1')$$
$$\pm i\int d3\; V(1-3)g_3(123,1'2'3^+). \qquad (3b)$$

The equations are very general, i.e. they do not depend on the kind of averaging. Therefore boundary conditions are necessary to obtain solutions for a special kind of averaging. For example in the case of thermodynamical equilibrium we have the well-known KMS-condition /2/.

In nonequilibrium situations this condition is not valid. Up to now there exist two methods to determine nonequilibrium GF.

i) KANDANOFF and BAYM have performed an analytic continuation of the imaginary time GF to their real-time counterparts /2/.

ii) In order to deal directly with real-time GF KELDYSH /3/ introduced a diagrammatic perturbation scheme in the interaction picture.

In this lecture we would like to introduce a new method to calculate the nonequilibrium real-time GF directly from the equations of motion. In this case we use the condition of weakening of initial correlation introduced in nonequilibrium mechanics by BOGOLJUBOV /4/. In terms of GF this condition has the following form (for detailed considerations we refer to /5,6,7/).

$$\lim_{t' \to t \pm \varepsilon \to -\infty} g_2(12,1'2')\Big|_{\substack{t_1=t_2=t \\ t_1=t_2=t}} = g_1^{\gtrless}(11')g_1^{\gtrless}(22') \pm g_1^{\gtrless}(12')g_1^{\gtrless}(21') \ . \qquad (4)$$

This condition is valid for situations in which the correlation time is much smaller than the time interval between two collisions.

In order to demonstrate this method, let us derive the Kadanoff-Baym equations for nonequilibrium GF. Our aim is to decouple the MS-hierarchy (3) by introducing a self-energy Σ in the first equation. This function shall have the property

$$\Sigma(1,1') = \Sigma^{HF}(1,1') + \theta(t_1-t_1')\Sigma^{>}(1,1') + \theta(t_1-t_1)\Sigma^{<}(1,1') \ . \qquad (5)$$

The Bogoljubov-condition (4) implies that

$$\lim_{t_1 \to -\infty} \pm i\int d2\ V(1-2)g_2(12,1'2^+)\Big|_{t_1'=t_1^{\pm}}$$
$$= \pm i\int d2\ V(1-2)\left[g_1(11')g_1(22^+) \pm g_1(12^+)g_1(21')\right]_{t_1'=t_1^{\pm}}$$
$$= \int d2\ \Sigma^{HF}(12)g_1(21')\Big|_{t_1'=t_1^{\pm}} \ , \qquad (6)$$

where $\Sigma^{HF}(11') = \Sigma^{HF}(\vec{r}_1,\vec{r}_{1'},t)\,\delta(t-t')$ is the Hartree-Fock self-energy.

An equation fulfilling this condition is given by

$$\left(i\frac{\partial}{\partial t_1} + \frac{\nabla_1^2}{2m}\right)g_1(1,1') = \delta(1-1') + \int_{-\infty}^{\infty} dt_2 \int d^3r_2 \left\{\Sigma(12)g_1(21') - \Sigma^{<}(12)g_1^{>}(21')\right\}. \qquad (7)$$

It is easy to get from this equation the corresponding equations for the correlation functions $g_1^{\gtrless}$ often called "Kadanoff-Baym equations" /2/.

Of course the decoupling of the hierarchy is only formal, i.e. in practice we need certain approximation for the self-energy or the two-particle GF, respectively. Starting from the Martin-Schwinger equation, it is possible to derive a nonequilibrium cluster expansion for the two-particle GF. Under the restriction to three-particle collisions we arrived at

$$g_2(12,1'2') = g_2^L(12,1'2') \pm \int d3d\bar{3}\ g_1^{-1}(3,\bar{3})\{g_3^L(12\bar{3},1'2'3^+)$$
$$-g_2^L(12,1'2')g_1(\bar{3},3^+)\Big|_{exch.}\} - g_2^{SE}(12,1'2') \ , \qquad (8)$$

where g_2^L and g_3^L are the solutions of the two- and three-particle ladder equation, and g_2^{SE} contains the self-energy contributions to the two-particle term, which are of the order of three-particle events.

The function g_2^L is given by

$$g_2^L(12,1'2')\Big|_{\substack{t_1=t_2=t \\ t_1=t_2=t}} = g_1(11')g_1(22')\Big|_{exch.} + i\int_{-\infty}^{\infty} d\bar{1}d\bar{2}\ V(\bar{1}-\bar{2})$$
$$*\{g_1(1\bar{1})g_1(2\bar{2})g_2^L(\bar{1}\bar{2},1'2') - g_1^{<}(1\bar{1})g_1^{<}(\bar{2}2)g_2^{>}(1\bar{2}1'2')\} \ . \qquad (9)$$

If we introduce generalized Moeller operators (the coordinates are suppressed in the following)

$$\Omega^R_{12}(t,t') = 1\,\delta(t-t')+i\,g^R_{12}(t,t')\,V_{12};\quad g^R_{12}=g_{12}-g^<_{12}\,, \tag{10}$$

we find for the correlation function $g^{\gtrless}_{12}$

$$g^{\gtrless}_{12}(t,t') = \int_{-\infty}^{\infty} d\bar{t}_1 d\bar{t}_2 \Omega^R(t\bar{t}_1) g^{\gtrless}_1(\bar{t}_1\bar{t}_2) g^{\gtrless}_1(\bar{t}_1\bar{t}_2) \Omega^A(\bar{t}_2 t')\,. \tag{11}$$

Before we deal with the three-particle contribution to the cluster expansion, we want to consider a more general situation in which the formation of bound states is possible in the system (for details we refer to the dissertation B of one of us (M.S.) /7/). In this case a generalization of the boundary condition is necessary. Obviously it is

$$\lim_{t'\to t\to-\infty} g^{\gtrless}_{12}(t,t') = g^{\gtrless\,\mathrm{bound}}_{12} + g^{\gtrless\,\mathrm{scattering}}_{12}, \quad \lim_{t'\to t\to-\infty} g^{\gtrless\,\mathrm{scatt.}}_{12} = g^{\gtrless}_1 g^{\gtrless}_2. \tag{12}$$

The corresponding condition for $g^{\gtrless}_{123}$ which takes into account the existence of two-particle subclusters is

$$\lim_{t'\to t\to-\infty} g^{\gtrless}_{123}(t,t') = g^{\gtrless}_1 g^{\gtrless}_2 g^{\gtrless}_3 + g^{\gtrless\,\mathrm{bound}}_{23} g^{\gtrless}_1 + g^{\gtrless\,\mathrm{bound}}_{13} g^{\gtrless}_2 + g^{\gtrless\,\mathrm{bound}}_{12} g^{\gtrless}_3 = \sum_{k=0}^{3} G^{k\,\gtrless}_{123}\,, \tag{13}$$

with

$$G^0_{123} = g_1 g_2 g_3,\quad G^1_{123} = g^{\mathrm{bound}}_{23} g_1,\ \text{etc.} \tag{14}$$

Then the solution of the three-particle ladder equation can be written down as

$$g^{\lessgtr}_{123}(t,t') = \sum_{k=0}^{3} \int_{-\infty}^{\infty} d\bar{t}_1 d\bar{t}_2\, \Omega^R_k(t,\bar{t}_1) G^{k\,\gtrless}_{123}(\bar{t}_1,\bar{t}_2) \Omega^A_k(t_2,t')\,; \tag{15}$$

here $\Omega^R_k(t,t') = 1\delta(t-t')+i^2 g^R_{123}(t,t')\,V^k_{123}$ is a generalization of the Moeller operators of the usual multi-channel scattering theory. Introducing the cluster expansion into the first hierarchy-equation and assuming very simple spectral properties (quasi-particle approximation) we obtain a kinetic equation for the one-particle distribution function. Because of condition (13), this equation is not closed, there is needed a second equation for the distribution of bound particles. If we assume a homogeneous system the generalized Boltzmann equation reads

$$\frac{\partial}{\partial T} f(p_1 T) = I_2 + I_3 - I_{SE}\,, \tag{16a}$$

$$\frac{\partial}{\partial T} f_{n_{12}}(P_{12},T) = I_{n_{12}}\,, \tag{16b}$$

with I_2 and I_3 being the two- and three-particle collision integrals, respectively. The term I_{SE} compensates for the successive binary collisions in I_3 and ensures the conservation of the energy in binary collision approximation. For example, I_3 is given by

$$I_3(p_1,T) = \frac{1}{2V} \sum_{\bar{k}}\!\!\!\!\!\int d\bar{p} \int \frac{dp_2}{(2\pi)^3} \frac{dp_3}{(2\pi)^3} \left|\langle p_1 p_2 p_3 | T^{0\bar{k}}_{123}(E^0_{123}+i\varepsilon) | \bar{k}\bar{p}\rangle\right|^2$$
$$*2\pi\,\delta(E^0_{123}-E^{\bar{k}}_{123})\{\bar{f}_{\bar{k}} - f_1 f_2 f_3\}$$

$$+ \frac{1}{2V} \sum_{\bar{k}} \int d\bar{p} \sum \frac{dP_{23}}{(2\pi)^3} |\langle p_1 P_{23} n_{23} | T^{1\bar{k}}_{123}(E^1_{123}+i\varepsilon) | \bar{k}\bar{p} \rangle|^2$$

$$*2\pi\delta(E^1_{123}-E^{\bar{k}}_{123})\{\bar{f}_{\bar{k}}-f_{n_{23}}f_1\} \; ; \qquad (17)$$

here $f_{k=0} = f_1 f_2 f_3$, $f_{k=1} = f_1 f_{n_{23}}$ etc.

The T-matrices describe the relevant processes (scattering, ionization, recombination, rearrangement). The system of kinetic equations (16) is the basis of a quantum kinetic description of chemical reacting systems (see also /8,9/).

References

1. P.C. Martin, J. Schwinger: Phys.Rev. 115, 1342 (1959)
2. L.P. Kadanoff, G. Baym: Quantum Statistical Mechanics (Benjamin, New York 1962)
3. L. Keldysh: Zh.Eksp.Teor.Fiz. 46, 1515 (1964)
4. N.N. Bogoljubov: Problems of Dynamical Theory in Statistical Mechanics (Gostechisd., Moscow 1946)
5. D. Kremp, M. Schlanges, Th. Bornath: J.Stat.Phys. 41, 661 (1985)
6. M. Schlanges, Th. Bornath, D. Kremp: Wiss.Zeitschrift WPU Rostock, NR 34, 5 (1985)
7. M. Schlanges: Diss. B, Rostock 1985
8. Yu.L. Klimontovich, D. Kremp: Physica 109A, 517 (1981)
9. J.A. McLennan: J.Stat.Phys. 28, 521 (1982)

Transport Phenomena in Many-Particle Systems and the Quantum Statistical Approach to Nonequilibrium Thermodynamics

G. Röpke

Wilhelm-Pieck-Universität Rostock, Sektion Physik, Universitätsplatz 3, DDR-2500 Rostock, GDR

A many-particle system described by the Hamiltonian

$$H_s = \sum_k E(k) a_k^+ a_k + \frac{1}{2} \sum_{kpq} V(k,p,q) a_{k+q}^+ a_{p-q}^+ a_p a_k \qquad (1)$$

is considered; k denotes the momentum and further internal quantum numbers as spin, isospin etc. of a single particle state, $E(k) = \hbar^2 * k^2/2m_k$; a_k^+, a_k are the Fermionic creation and annihilation operators, respectively. To be specific, let us consider a Coulomb system consisting of electrons and protons so that $V(k,p,q) = e_k e_p/\varepsilon_0 q^2$. As a consequence of this interaction, bound states such as Hydrogen atoms and molecules may be formed, and in the density-temperature plane different states occur, like the fully ionized plasma, the partially ionized plasma, atomic and molecular Hydrogen gas, solid, liquid, and metallic Hydrogen, cf. Fig. 1.

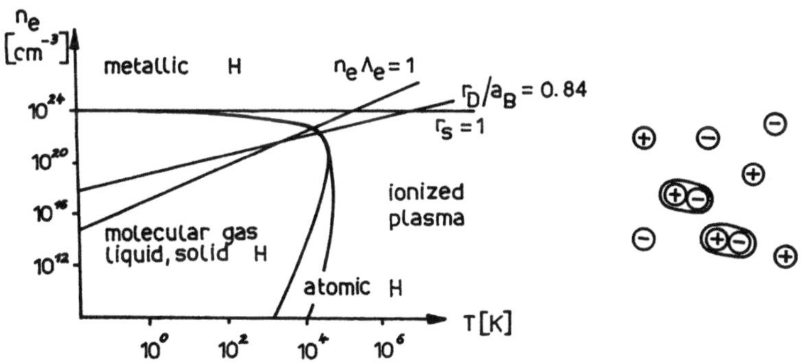

Fig. 1 Phase diagram of Hydrogen, schematically

The thermodynamics of such a many-particle system for arbitrary degeneration is investigated quite well by using the technique of thermodynamic Green functions and diagram representation. Many-particle effects as dynamical screening, self-energy, bound state formation, Pauli blocking and Mott-effect are treated in a systematic way. Rigorous results are known in some limiting cases of a Coulomb system, and interpolation formulae for the equations of state have been constructed for the entire density-temperature plane, indicating the occurrence of possible phase transition. A review of the quantum statistical approach to Coulomb systems is found in the monographs /1/.

On the other hand, the treatment of transport properties as, e.g. the static conductivity σ has not been elaborated in an adequate level until now. By reason of dynamical screening, e.g., there are no rigorous results for σ in the limiting cases. Different formulae are known which work quite well in the corresponding regions of the density-temperature plane. For instance, the electrical conductivity of the fully ionized plasma is well described by the Spitzer formula, the Ziman formula is applied to liquid metals, and the Brooks-Herring formula to semiconductors, see also /2/.

A systematic quantum statistical approach to the transport coefficients of a Coulomb system which works in the whole density-temperature plane was elaborated during the last years, cf. ref. /3/. A very general approach to nonequilibrium has been used which combines macroscopical thermodynamical description and microscopical dynamics. This approach along the lines of the Zubarev method /4/ allows us to derive generalized kinetic equations as well as linear response theory. In principle, this formalism describes also systems far from equilibrium. Nonlinear macroscopic behaviour may be described with the help of an appropriate macroscopic reference state, but these problems were not considered in detail until now within this formalism.

2. The Nonequilibrium State of a Many-Particle System

Hamiltonian dynamics is not sufficient to describe a system in nonequilibrium, and all microscopic approaches to nonequilibrium make more or less clearly additional assumptions on the time evolution of the considered system which violate the Hamiltonian dynamics. This becomes most obvious if the production of entropy is considered. A familiar form to perturb the Hamiltonian dynamics of a system is the so-called principle of weakening of initial correlations. As shown by ZUBAREV /4/, the most natural way to formulate this principle in order to guarantee the validity of thermodynamics in equilibrium and the macroscopic theory of irreversible processes reads

Hamiltonian dynamics / Liouville-von Neumann-equation ↘   Principle of weakening of initial correlation ↙

$$\frac{\partial}{\partial t}\varrho(t) + \frac{i}{\hbar}[H(t), \varrho(t)] = -\varepsilon(\varrho(t) - \varrho_{rel}(t)) \tag{2}$$

(equation of motion for the statistical operator $\varrho(t)$), where the relevant statistical operator

$$\varrho_{rel}(t) = Z_{rel}^{-1} \exp\left\{-\sum_n F_n(t) B_n\right\} \tag{3}$$

represents a generalized Gibbs state with respect to a given set $\{B_n\}$ of relevant observables. The thermodynamic parameters F_n are determined by the condition that the given mean values $\langle B_n\rangle^t \equiv \mathrm{Tr}\,\{\varrho(t)B_n\}$ are correctly reproduced with $\varrho_{rel}(t)$:

$$\langle B_n\rangle^t \equiv \mathrm{Tr}\,\{\varrho(t)B_n\} = \mathrm{Tr}\,\{\varrho_{rel}(t)B\} \equiv \langle B_n\rangle^t_{rel}\,. \tag{4}$$

The limit $\varepsilon \to 0$ is performed after the thermodynamic limit.

The principle of weakening of initial correlations (r.h.s. of (2)) which may be considered as a basic principle in nonequilibrium statistical mechanics acts as an (infinitesimal) perturbation to the Liouville-von Neumann-equation (l.h.s. of (2)) and may be formulated with the help of the Abel theorem. The generalized Gibbs state (3) for $\varrho_{rel}(t)$ corresponds to a maximum of entropy defined in nonequilibrium by

$$S(t) = - k_B \langle \ln \varrho_{rel}(t) \rangle^t = - k_B \langle \ln \varrho_{rel}(t) \rangle^t_{rel} \quad , \tag{5}$$

with given constraints $\langle B_n \rangle^t$. Usually, for a closed system, the time evolution of the relevant observables $\langle B_n \rangle^t$ should follow from the equation of motion (2) so that we have a self-consistent problem in determining $F_n(t)$. However, for an open system we can also consider $\langle B_n \rangle^t$ or $F_n(t)$ as given external control parameters with prescribed time dependence. For instance, the formation of structures in nonlinear nonequilibrium systems can be represented by the evolution of the macroobservables $\langle B_n \rangle^t$.

Of course, there are questions with respect to the magnitude of ε in (2) which should be considered as a finite quantity during the calculation, and with respect to the choice of the set $\{B_n\}$ of macroobservables which determine $\varrho_{rel}(t)$. We give three examples with respect to the many-particle system (1) in Table 1.

Table 1 Different descriptions of the nonequilibrium state as consequence of the corresponding choice of macroobservables B_n

Kubo formalism	hydrodynamic equations	kinetic equations
$\{B_n\} = \{H_s, N\}$	$\{B_n\} = \{h(\vec r), n(\vec r), \vec P(\vec r)\}$	$\{B_n\} = \{n(\vec k), n(\alpha P)\}$
(conserved quantities)	(densities of conserved quantities)	(occupation numbers of asymptotic free or bound states)
$\varrho_{rel} = \varrho_0 = Z_0^{-1} \exp(-\beta * (H_s - \mu N))$	$\vec P = \sum_{i,a} p_{i,a} = \sum_k \hbar \vec k a_k^+ a_k$	

We will apply this formalism to the case of an open many-particle system under influence of an external field $\vec E$,

$$H = H_S + H_F = H_S - \vec E \sum_{a,i} e_a \vec r_i = H_S - \vec E \sum_{kk'} e_k i\hbar \frac{\partial}{\partial \vec k} \delta(\vec k - \vec k') a_k^+ a_{k'} \quad , \tag{6}$$

in contact with a bath, see Fig. 2, in the stationary state (for the time-dependent case cf. ref. /5/).

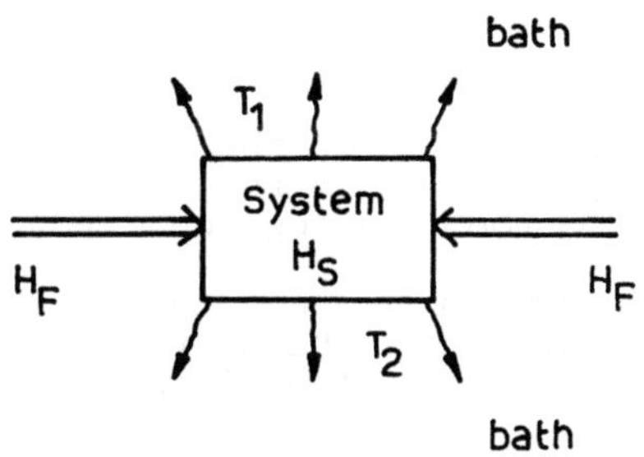

Fig. 2 Open system under the influence of an external field in the stationary state

The transport coefficients relate the internal currents to the generalized forces as electrical field and temperature gradient. For instance, the static electrical conductivity $\sigma = \sigma(\vec{q}=0, \omega=0)$ is defined according to

$$\vec{j}_{el} = \frac{1}{\Omega} \mathrm{Tr}\{\varrho \sum_k \frac{e_k}{m_k} \hbar \vec{k} a_k^+ a_k\} = \frac{1}{\Omega} \langle \frac{e}{m} \vec{p} \rangle = \sigma \vec{E} , \quad (7)$$

with Ω -system volume. The statistical operator ϱ follows from the formal solution of (2),

$$\varrho(t) = \varepsilon \int_{-\infty}^{t} dt' e^{\varepsilon(t'-t)} U(t,t') \varrho_{rel}(t') U^{-1}(t,t') \quad (8)$$

and in the stationary case after integration by parts

$$\varrho = \varrho_{rel} - \int_{-\infty}^{0} dt\, e^{\varepsilon t}\, e^{\frac{i}{\hbar}Ht} \frac{i}{\hbar} [H, \varrho_{rel}]\, e^{-\frac{i}{\hbar}Ht} . \quad (9)$$

3. Kinetic equations

The ordinary Boltzmann equation is immediately derived within our scheme under the following assumptions:
(i) the set $\{B_n\}$ of relevant observables is given by the occupation numbers n_k of the single particle states;
(ii) no retardation, $\varrho_{rel}(t') \approx \varrho_{rel}(t)$;
(iii) the two-particle T-matrix (binary collision approximation).

The following generalizations are possible:
(i') extension of the set of relevant observables by including, e.g., the occupation number $n_{\alpha P}$ of bound states with internal quantum number α and total momentum P or by introducing the quasiparticle picture, which accounts for long-living correlations /3,6/.
(ii') The treatment of the non-stationary case /5/ gives the dispersion $\sigma(q,\omega)$ of transport coefficients.
(iii') The correlation functions in the state described by ϱ_{rel} should be evaluated using many-particle techniques /7/.

For details of the derivation of kinetic equations from the general equation of motion (2) see /8/. Note that we are able also to describe systems far from equilibrium, because the relevant statistical operator $\varrho_{rel}(t)$ (3) follows from a variational principle, and the thermodynamic parameters $F_n(t)$ must not reflect small deviations from equilibrium. Generalizations of kinetic equations are also derived from the Green function approach to many-particle systems /9/ but the relation to nonequilibrium thermodynamics was not considered.

4. Linear Response Theory

Within the linear response theory, explicit results for the transport coefficients are obtained in the form of equilibrium correlation functions which allow for a many-particle treatment using the formalism of thermodynamic Green functions.

Treating the many-particle system (1) by perturbation theory with respect to the interaction, we can employ some ideas which are expressed by performing partial summations, see the next Section. One guiding idea is the chemical picture:
Bound states are treated on the same level as free particles.

In a dense system, we have to look for long-living correlations (nearly conserved observables) to define $\varrho_{rel}(t)$ and we have to con-

Table 2 Choice of relevant observables $\{B_n\}$ for a partially ionized plasma without large-scale fluctuations

$\{B_n\}$	low density		dense systems
occupation numbers $\{n_\nu\}$ for	free particle states $\vert\nu\rangle = \vert k\rangle$	→	quasiparticle states
	bound states $\vert\nu\rangle = \vert nP\rangle$	→	effective in-medium two-particle states

sider quasiparticle states which take into account self-energy effects; in general, see Table 2, occupation numbers of further quasiparticle state as photons, phonons, etc. should be considered. Note, that in the case that a thermodynamic instability arises (formation of dissipative structures), the initial conditions are long living, and the relevant statistical operator $\varrho_{rel}(t)$ has to describe the preparation or small perturbations during instability.

The thermodynamical parameters F_ν of

$$\varrho_{rel} = Z_{rel}^{-1} \exp\left\{-\beta(H_S+\mu N) - \sum_\nu F_\nu \, n_\nu\right\} \tag{10}$$

have the meaning of a chemical potential. On the one hand, they are related to the distribution functions f_ν of free particles and of bound states,

$$f_\nu = \langle n_\nu\rangle = \langle n_\nu\rangle_{rel} = f_\nu^{eq} + \sum_{\nu'} F_{\nu'} \, (n_{\nu'} | n_\nu) \, , \tag{11}$$

if we expand for small deviations from equilibrium, on the other hand they are determined by the generalized linear Boltzmann equations

$$E\left\{(P|n_\nu)+\langle P;\dot{n}_\nu\rangle\right\} = \sum_{\nu'} F_{\nu'} \left\{(\dot{n}_{\nu'}|n_\nu)+\langle\dot{n}_{\nu'};\dot{n}_\nu\rangle\right\} \, , \tag{12}$$

with the notations

$$\langle A;B\rangle = \int_{-\infty}^{0} dt \, e^{\varepsilon t} \, (A(t)|B), \quad (A|B) = \int_0^\beta d\tau \, \mathrm{Tr}\left\{\varrho_0 A(-i\hbar\tau)B\right\} \tag{13}$$

Equation (12) is obtained from the conditions of stationarity

$$\frac{d}{dt}\langle n_\nu\rangle = \mathrm{Tr}\left\{\varrho \, \frac{i}{\hbar}[H,n_\nu]\right\} = 0 \, , \tag{14}$$

if ϱ, eq. (9), is expanded within the linear response approach up to first order with respect to E, F_ν. Because of symmetry and positive definiteness of the generalized Onsager coefficients $(\dot{n}_{\nu'}|n_\nu)+\langle\dot{n}_{\nu'};\dot{n}_\nu\rangle$, a variational principle /5/ (minimum production of entropy) can be applied to solve (12) and to find explicit expressions for the transport coefficients, see /10/.

5. Perturbative Expansion of Correlation Functions and Diagram Representation

After the transport coefficients are related to equilibrium correlation functions (13) we can apply standard many-particle techniques, e.g., the method of thermodynamic Green functions:

$$\langle A;B\rangle = \int_{-\infty}^{0} dt \, e^{\varepsilon t} \int_0^\beta d\tau \int_{-\infty}^{\infty} \frac{d\omega}{2\pi i} \, \frac{1}{e^{\beta\omega}-1} \cdot e^{i\omega(t-i\hbar\tau)/\hbar} \left[G_{AB}(\omega+i\varepsilon) - G_{AB}(\omega-i\varepsilon)\right] . \tag{15}$$

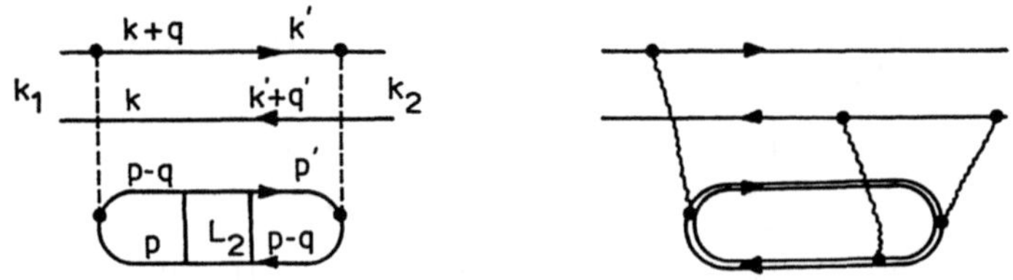

Fig. 3 Polarization approximation and dynamically screened 2nd Born approximation

The perturbative expansion of Green functions can be represented by diagrams, and physical arguments as dynamical screening, polarization function, self-energy, bound-state formation and the chemical picture are represented by partial summations of diagrams. We give two examples for approximations of the collision term $\langle \dot n_{k_1} | \dot n_{k_2}\rangle$, the polarization approximation and the dynamically screened 2nd Born approximation, see Fig. 3.

The results are /6/ (f, f_3 denote Fermi functions, n_B the Bose function)

$$\langle \dot n_{k_1}; \dot n_{k_2}\rangle^{pol} = \frac{2\beta}{\hbar}\sum_{kk'} e_k^2\, V(k'-k)\int d\omega\ \mathrm{Im}\,\varepsilon^{-1}(k'-k,\omega)(1+n_B(\omega))\, f_k[1-f_{k'}]$$
$$\cdot\,\delta(\omega - E_k + E_{k'})\,\delta_{k_1,k}\,(\delta_{k_2,k'} - \delta_{k_2,k_1}), \qquad (16)$$

which reproduces in dependence of the approximation for $\varepsilon(q,\omega)$ the linearized Landau collision term, the linearized Lenard-Balescu collision term or the contribution of collisions with bound states;

$$\langle \dot n_{k_1}; \dot n_{k_2}\rangle^{2\ \mathrm{Born}} = \frac{4\pi\beta}{\hbar}\sum_{kk'nn'PP'} f_3(E_k+E_{nP})\,\delta(E_k+E_{nP}-E_{k'}-E_{n'P'})$$
$$*\ |V^{(2)}(knP,k'n'P')|^2\,\delta_{k_1,k}\,(\delta_{k_2,k'}-\delta_{k_2,k_1})\,\delta_{k+P,k'+P'}\ , \qquad (17)$$

where $V^{(2)}$ contains the contribution of the dynamically screened polarization potential

$$V^{(2)}_{PP}(knP,k'n'P') = \sum_{k''n''P''}\frac{\delta_{k+P,k''+P''}}{E_k+E_{nP}-E_{k''}-E_{n''P''}}\,M(knP,k''n''P'')M(k''n''P'',k'n'P')\ ,$$

$$M(k,n,P,k+q,n',P-q) = e_k\int d^3p\,\Psi_n(p)\Big\{e_a\Psi_{n'}\big(p+\frac{m_b}{m_a+m_b}q\big)+e_b\Psi_{n'}\big(p-\frac{m_a}{m_a+m_b}q\big)\Big\}$$

$$*\ V(q)\Big\{1-\int\frac{d\omega}{\pi}\,\mathrm{Im}\,\varepsilon^{-1}(q,\omega^+)\Big[\frac{n_B(\omega)}{E_k+\omega-E_{k+q}}+\frac{1+n_B(\omega)}{E_{nP}-\omega-E_{n'P-q}}\Big]\Big\}\ . \qquad (18)$$

The evaluation of the conductivity σ of a partially ionized plasma in the corresponding approximation is given in ref. /10/.

6. Convergence of the Perturbative Expansion of Correlation Functions

As we have shown that the results of ordinary kinetic theory are reproduced within our approach to nonequilibrium thermodynamics, and, that new results for dense strongly coupled systems can be derived, we will point out some problems in the present approach. The first one is what happens if we take a reduced set of relevant observables $\{B_n\}$. The second one refers to the convergence of the perturbation expansion of the correlation functions which has to be performed for finite values of ε, but in the final results divergent terms may occur if the limit $\varepsilon \to 0$ is taken. Both problems are tightly connected, see ref. /5/.

Let us consider the one-component fully ionized plasma, with $\vec{j}_{el} = (e/m\Omega) Tr\{\varrho \vec{P}\}$, $\vec{P} = \sum_k \hbar \vec{k}\, a_k^+ a_k$. With the choice $\{B_n\} = \{H_s, N, \vec{P}\}$ for the set of relevant parameters we have after linearization of ϱ with respect to E, $\langle P \rangle$

$$\sigma = \frac{e^2}{m^2\Omega} \frac{(P|P)\{(P|P)+\langle P;\dot{P}\rangle\}}{\langle \dot{P};\dot{P}\rangle} = \frac{e^2 n}{m} \frac{Q}{\langle \dot{P};\dot{P}\rangle} , \tag{19}$$

with $P = \vec{P}\,\vec{E}/|\vec{E}|$, $\dot{P} = \frac{i}{\hbar}[H_s, P]$. Using the Kubo identity

$$(P|P) = (m\dot{R}_x|P_x) = \frac{i}{\hbar} Tr\{\varrho_0 [P_x, m\, R_x]\} = mN \tag{20}$$

and integration by parts $\langle \dot{P};\dot{P}\rangle = (P|\dot{P}) - \varepsilon \langle P;\dot{P}\rangle$, $\langle \dot{P};P\rangle = -\langle P;\dot{P}\rangle = (P|P) - \varepsilon\langle P;P\rangle$, we find the Kubo formula

$$\sigma = \frac{e^2 n}{m} \frac{\langle P;P\rangle}{\langle \dot{P};P\rangle} = \frac{e^2}{m^2\Omega} \frac{\langle P;P\rangle}{1-\varepsilon \frac{1}{mN}\langle P;P\rangle} = \frac{e^2}{m^2\Omega}\langle P;P\rangle , \tag{21}$$

if the correlation function $\langle P;P\rangle$ remains finite in the limit $\varepsilon \to 0$.

The Kubo formula which may be derived with the reduced set of observables $\{B_n\} = \{H_s, N\}$ gives no convergent perturbative expansion with respect to the interaction; in lowest order we have

$$\langle P;P\rangle = \frac{mN}{\varepsilon} . \tag{22}$$

Results equivalent ot the treatment of the Boltzmann equation are obtained after summing up an infinite series of divergent terms $\sim \varepsilon^{-n}$, see ref. /11/.

In (19), a perturbation expansion with respect to the interaction can be performed where no divergency arises in the lowest order, and the well-known Ziman formula (Born approximation) is reproduced. However, results equivalent to the Boltzmann equation are obtained only after partial summation of divergent terms $\sim \varepsilon^{-n}$ in the higher order contributions, see ref. /12/. This partial summation is not necessary if we work with the extended set of relevant observables including the occupation numbers of single particle states.

As an example, let us consider the expression Q of (19) up to the first order with respect to the screened potential (Montroll-Ward approximation) in the non-degenerate limit, $f \ll 1$

$$Q = mN + \frac{\hbar^2 e^4}{3} \sum_{kpq} \vec{k}(\vec{k}+\vec{q}) f(E_p) V(q) V^s(q) \frac{1}{\Delta^2} \{\beta f(E_k - \Delta)$$

$$+ \beta f(E_k) + \frac{2\varepsilon^2}{\varepsilon^2 + \Delta^2} \frac{1}{\Delta} [f(E_k) - f(E_k - \Delta)]\} \tag{23}$$

with $\Delta = E_k + E_p - E_{k+q} - E_{p-q}$. The result

$$Q - mN = -mN \beta e^2 \varkappa/6 - 2\hbar^2\, mNe^2 \varkappa/\beta\varepsilon^2 \pm \dots \tag{24}$$

describes the Debye-Onsager relaxation effect, see /13/.

7. Stochastic Forces

The expression (19) relates the inverse transport coefficient, the resistivity, to the correlation function of real forces P in the system, divided by the denominator Q. After simple algebraic manipulations we obtain

$$\frac{1}{\sigma} = \frac{1}{e^2 nN} \langle R;R \rangle \quad ; \quad R = \dot{P} - \frac{\langle \dot{P};P \rangle}{\langle P;P \rangle} P \,. \qquad (25)$$

This relation is in analogy to the fluctuation-dissipation theorem of second type which relates the resistivity to the fluctuations of stochastic forces within a Langevin approach.

Of course, R is not a stochastic process but is determined by the Hamiltonian dynamics, cf. (19). However, we can make the relation between our approach to nonequilibrium based on (2) and the Langevin approach more closely. For this we use the Heisenberg picture instead of the Schrödinger picture, and find

$$\frac{d}{dt} \langle A \rangle = \mathrm{Tr}\left\{\frac{i}{\hbar} [\varrho(t),H] A\right\} - \varepsilon [\mathrm{Tr}\{\varrho A\} - \mathrm{Tr}\{\varrho_{rel} A\}], \qquad (26)$$

so that

$$\frac{d}{dt} A = \frac{i}{\hbar} [H,A] - \varepsilon (A - \mathrm{Tr}\{\varrho_{rel} A\}) \,, \qquad (27)$$

i.e. there occurs a term additional to the Hamiltonian dynamics. Because of (4), only the set of relevant observables $\{B_n\}$ evolves in the average due to the Hamiltonian dynamics, the other observables follow a time evolution with an additional contribution similar to a relaxation term. Now, we construct a Langevin equation, according to the idea that any relaxation is connected with a stochastic process, and find

$$dA = \frac{i}{\hbar} [H,A] \, dt - \varepsilon (A - A_{rel}) \, dt + dw(t) \,, \qquad (28)$$

where dw(t) is given, for instance, by a Wiener process.

8. Conclusions

We showed that a very general approach to nonequilibrium is found if the Hamiltonian dynamics is changed by a relevant term which expresses the principle of weakening of initial correlation and introduces the concept of entropy into statistical mechanics. This approach works far from equilibrium as well as near equilibrium where the standard many-particle techniques are employed to take into account many-particle effects in evaluating the transport coefficients. Different choices of the set of relevant observables may be compared in linear response theory with the principle of minimum entropy production if the state of the system (distribution function) is varied, extending the class of functions. The relation to stochastic processes is pointed out.

References

1. W.D. Kraeft, D. Kremp, W. Ebeling, G. Röpke: Quantum Statistics of Coulomb Systems (Akademie-Verlag und Plenum Press, Berlin 1986)

W. Ebeling, W.D. Kraeft, D. Kremp: in: Ergebnisse der Plasmaphysik und Gaselektronik, Vol. 5, Akademie-Verlag, Berlin 1976
2. G. Röpke, F. Höhne: phys.stat.sol.(b) 107, 603 (1981)
3. W. Ebeling et al.: Transport Properties of Dense Plasmas (Birkhäuser, Basel 1984)
G. Röpke: Teor.mat.Fiz. 46, 279 (1981)
4. D.N. Zubarev: Nonequilibrium Statistical Thermodynamics (Plenum Press, New York London 1974)
D.N. Zubarev: in: Itogi Nauki i Tekhniki, Vol. 15, VINITI, Moscow 1980, p. 131
5. V. Christoph, G. Röpke: phys.stat.sol.(b) 131, 11 (1985)
6. G. Röpke: Physica 121A, 92 (1983)
7. A. Fetter, J.D. Walecka: Quantum Theory of Many-Particle Systems (McGraw-Hill, New York 1971)
8. G. Röpke: ZIE preprint 1986
9. L.P. Kadanoff, G. Baym: Quantum Statistical Mechanics (Benjamin, New York 1962)
D. Kremp, M. Schlanges, Th. Bornath: J.Stat.Phys., to appear; ZIE preprint 1986
10. F.E. Höhne, R. Redmer, G. Röpke, H. Wegener: Physica 128A, 643 (1984)
11. G.V. Chester, A. Thellung: Proc.Phys.Soc. 73, 745 (1959)
12. M. Huberman, G.V. Chester: Adv.Phys. 24, 489 (1975)
13. W. Ebeling, G. Röpke: Ann.Physik (Leipzig) 36, 429 (1979)

Part III

Dissipative Structures

Dynamics of Interfaces in Random Media

A. Engel, W. Ebeling, R. Feistel, and L. Schimansky-Geier

Humboldt-Universität Berlin, Sektion Physik, Invalidenstraße 42, DDR-1040 Berlin, GDR

1. Introduction

One of the essential consequences of non-linear equations of motion is the possibility of several stable stationary states /1/. As a result, for a d-dimension system many configurations exist, where regions in which the system is in one of its stable states are separated by thin transition layers. As the non-linearity increases, the widths of these layers decrease, so that they can be described as (d-1)-dimensional hypersurfaces. The time evolution of the system is then determined by the dynamics of these phase-separating interfaces. Situations of this type arise in a variety of physical systems. Well-known examples are equilibrium phase transitions of first order as, e.g., liquid-vapour systems /2/. Of great practical importance are furthermore systems quenched into the phase region of distinct multistability /3-6/. After a very quick local relaxation process, a complicated pattern of domains separated by interfaces emerge. These domains coarsen with time, as can be verified experimentally by scattering techniques. Since the coarsening process is determined by the dynamics of the existing interfaces, comparison between theory and experiment becomes possible /5/. Of interest for practical purpose are in particular metallurgical systems as, e.g., binary alloys /3/.

Another interesting example is the ballast resistor /7-9/, a divice used already for a long time for stabilizing an electric current. A current-conducting wire in a gas tube can separate under appropriate conditions into a high-temperature (i.e. high-resistance) and low-temperature (i.e. low-resistance) region connected by a thin transition layer similar to the liquid-vapour-interface of a Van der Waals gas. If the voltage in the circuit changes, this transition layer is shifted, and hence the total resistance of the wire changes, thereby ensuring a constant current (as the liquid-vapour interface of a Van der Waals gas moves in order to ensure constant pressure if the volume is altered).

We mention also pumped superconductors /10/, turbulent fluids /11/, nonequilibrium chemical systems /12/ and charge-density waves in solids /13/.

A variety of other systems are discussed in the books /14/ and /5/.

For definiteness we consider here bistable systems with non-conserved scalar order parameter $n(r,t)$, $r \in R^d$. The equation of motion is of reaction-diffusion-type:

$$\partial_t n(q,t) = - \partial_n U(n(r,t)) + D \partial_x^2 n(r,t) . \qquad (1.1)$$

The potential U is assumed to have two deep minima at the stable homogeneous configurations $n(r,t) \equiv n_1$ and $n(r,t) \equiv n_3$, D denotes the diffusion coefficient. Dealing mostly with thermodynamically open systems, the question about the influence of fluctuations becomes important. Furthermore, the order parameter can interact with quenched disorder as represented, e.g., by impurities in solids or inhomogeneities in chemically reacting gels. To account for these influences we add a noise term to (1.1) and obtain

$$\partial_t n(r,t) = -\partial_n U(n) + D\partial_x^2 n(r,t) + H(r,t) . \tag{1.2}$$

For the random field H we assume generally Gaussian statistics given by

$$\overline{H(r,t)} \equiv 0 , \quad \overline{H(r,t)H(r't')} = S(|r-r'|,|t-t'|) . \tag{1.3}$$

$S(r,t) \sim T \cdot \delta(r)\,\delta(t)$ corresponds to thermal noise, $S(r,t) = S(r)$ to quenched disorder.

The paper is organized as follows. In section 2 we recall the deterministic dynamics of interfaces in bistable media as resulting from (1.1). Section 3 is devoted to the main problem, namely the influence of the noise term H(r,t) on this interface dynamics. As is well known, the fluctuations of the interface position are the most significant ones, since they are associated with the Goldstone-modes of spontaneously broken translational invariance /15-17/. They therefore combine to form a gapless excitation, whereas all other modes have a gap, which increases with increasing non-linearity. The random variations of the interface position discussed in section 3 are therefore the most significant consequences of the noise term in (1.3). Finally we give a short summary of the main results obtained in section 4.

2. Deterministic Interface Dynamics

We consider a system described by (1.1) with a bistable potential U(n). As is well known, (1.1) possesses a plane kink solution, i.e. introducing a coordinate system by $r = (x,z)$, $x \in R^{d-1}$, $z \in R$, there exists a solution of (1.1) of the form

$$n(r,t) = n_0(z-vt) .$$

$n_0(\xi)$, $\xi = z-vt$ therefore satisfies

$$0 = -\partial_n U(n_0) + v\,\partial_\xi n_0(\xi) + D\partial_x^2 n_0(\xi) . \tag{2.1}$$

$n_0(z-vt)$ interpolates between n_1 and n_3, the two stable states of the system, and the phase-separating interface $f(x) = v\cdot t$ moves along the positive z-axis with constant velocity v determined by $U(n_1) - U(n_3)$ (Fig. 1). A locally spherical interface can be described by the density profile $n(r,t) = n_0(r-R(t))$ whenever the radius R is large as compared with the interface thickness. Using the Laplacian in spherical coordinates, we find from (1.1)

$$0 = -\partial_n U(n_0) + \left[\frac{D(d-1)}{r} + \partial_t R(t)\right]\partial_r n_0 + D\,\partial_r^2 n_0 \tag{2.2}$$

and considering the interface position $r = R(t)$, comparison of (2.1) and (2.2) yields

$$\partial_t R(t) = v - \frac{D(d-1)}{r} . \tag{2.3}$$

(2.3) describes the effect of a local curvature on the interface velocity (Fig. 2). Generalizing (2.3) to arbitrary interface configurations one finds for the local, normal interface velocity $v_\perp$:

$$v_\perp = v + DK , \tag{2.4}$$

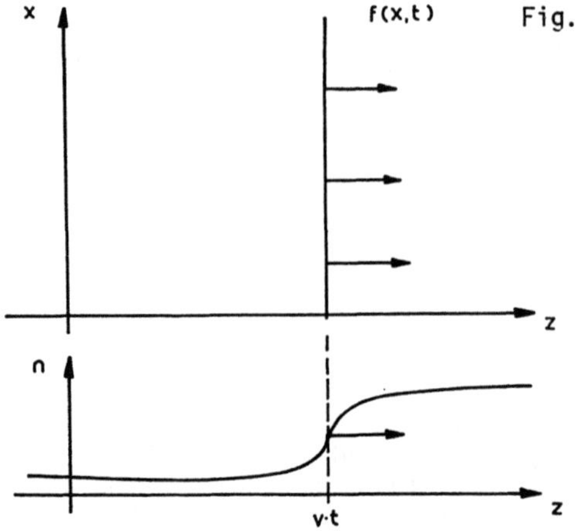

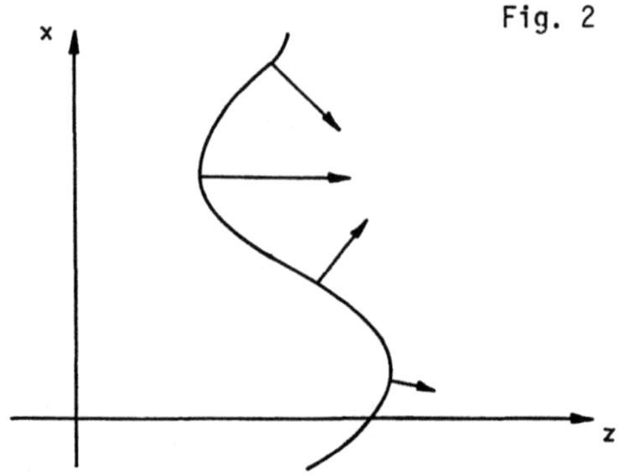

Fig. 1 Plane interface configuration $f(x,t) = vt$ and corresponding density profile $n_0(z-vt)$ in d=2 dimensions

Fig. 2 The normal velocity of a curved interface depends on the local mean curvature (cf. (2.4))

where K denotes the local mean curvature of the interface (/3,18/). We note that (2.4) can be derived systematically from (1.1) in the limit of strong non-linearity (/19,16,17/). Denoting the interface position by f(x,t) we have in general

$$v_\perp = \frac{\partial_t f(x,t)}{\sqrt{1+(\nabla f)^2}} \quad , \qquad K = \nabla \frac{\nabla f}{\sqrt{1+(\nabla f)^2}} \; . \tag{2.5}$$

However, we will restrict ourselves here to the case $|\Delta f| \ll 1$ so (2.4) and (2.5) simplify to

$$\partial_t f(x,t) = v + D\,\partial_x^2 f(x,t) \; . \tag{2.6}$$

(2.4) and (2.6) describe the deterministic kinetics of interfaces in bistable nonequilibrium systems as result of the interplay between a global contribution v related to the supersaturation $U(n_1) - U(n_3)$ and a local contribution $D\,\partial^2 f(x,t)$, which tries to flatten the interface.

3. Effects of Disorder

In order to study the influence of disorder on the interface dynamics we start with (1.2) rather than with (1.1) and obtain for $|\nabla f| \ll 1$ analogously to (2.6) the equation

$$\partial_t f(x,t) = v + D\,\partial_x^2 f(x,t) + h(x,f(x,t),t) \; , \tag{3.1}$$

where the random field h(x,z,t) is given by /20,21/

$$h(x,z,t) = \left[\int d\xi\, e^{\frac{v\xi}{D}} (\partial_\xi n_0)^2\right]^{-1} \int d\xi\, e^{\frac{v\xi}{D}} \partial_\xi n_0(\xi) H(x,z+\xi,t) . \tag{3.2}$$

Since h(x,z,t) depends linearly on H(r,t) its statistics is also Gaussian with zero mean and the correlation function

$$\overline{h(x,z,t)h(0,0,0)} = K(x,z,t) \tag{3.3}$$

can be obtained from (3.2) and (1.3).

In solving (3.1) we have first to treat the non-linear last term properly, and then to average the solution obtained over the realizations of the random field. Both tasks can be carried out most efficiently by using an iteration technique proposed by EFETOV and LARKIN in the context of charge-density waves /22/. The applicability of the method has been discussed in detail elsewhere /23/. We insert the ansatz

$$f(x,t) = v\, t + \sum_i f_i(x,t) \;;\; f_i = O(h^i) \tag{3.4}$$

into (3.1) and matching powers of h obtain the f_i one after another with the help of the Green function G(x,t) of the operator $(\partial_t - D\partial_x^2)$ given by its fourier transform

$$G(k,t) = \int d^{d-1}x\; e^{-ikx} G(x,t) = \theta(t) e^{-Dk^2 t} \;.$$

In particular we find

$$f_1(x,t) = \int dx'dt'\; G(x-x',t-t')h(x',vt',t') \tag{3.5}$$

and

$$f_2(x,t) = \int dx'dx''dt'dt''\; G(x-x',t-t') * \\ * \; G(x'-x'',t'-t'')h(x',vt',t')\frac{\partial h}{\partial z}(x'',vt'',t'') \;. \tag{3.6}$$

For higher orders a diagrammatic representation is appropriate /22/, however, we stick here to the first and second order results (3.5) and (3.6).

Physically one expects local deformations of an initially planar interface due to the random field, which will be counter-balanced by the surface tension term $D\partial_x^2 f$. A quantitative description of this roughening phenomenon is given by the so-called statistical width $\overline{(f-\bar{f})^2}$. From (3.5) we find to first order

$$\overline{(f-\bar{f})^2} \simeq \overline{f_1^2} = \int dx'dx''dt'dt''\; G(x-x'',t-t'') \\ * \; G(x-x',t-t')\; K(x'-x'',\; v(t'-t''),\; t'-t'') \;. \tag{3.7}$$

For quenched disorder, K(x,z,t) = K(x,z) the integrals can be performed after fourier transformation yielding

$$\overline{f_1^2} = \int \frac{dq}{2\pi} \; \frac{d^{d-1}k}{(2\pi)^{d-1}} \; \frac{K(q,k)}{D^2k^4+q^2v^2} \quad , \text{ where} \tag{3.8}$$

$$K(q,k) = \int d^{d-1}x\; dz\; e^{-ikx-iqz}\; K(x,z)$$

denotes the fourier transform of K(x,z).

For $d \leqslant 3$ the main contribution in (3.8) comes from long wavelength fluctuations. Denoting the length of the system in x-direction by L we therefore find

$$\overline{f_1^2} \sim \frac{K(0)}{v'D}\, L^{3-d} \;. \tag{3.9}$$

(3.9) implies

$$\overline{(\nabla f)^2} \sim \frac{K(0)}{D\ v} L^{1-d} ,$$

so the approximation (2.6) is justified a posteriori for sufficiently large interface velocity v. Furthermore we can identify the ansatz (2.4) as an expansion around the limit of a flat interface moving with infinite velocity.

For time-dependent disorder with short correlation time, $K(x,z,t) \sim K(x,z)\ \delta(t)$, we find from (3.7) after fourier transformation

$$\overline{f_1^2} = \frac{1}{2D} \int \frac{dq}{2\pi} \int \frac{d^{d-1}k}{(2\pi)^{d-1}} \frac{K(q,k)}{k^2} ,$$

$$\sim \frac{K(0)}{D\ \Delta} L^{3-d} , \tag{3.10}$$

where Δ denotes the spatial correlation length. This result depends not on v, i.e. the motion of the interface is irrelevant. We mention that in this case $\overline{(f-\overline{f})^2}$ is given exactly by $\overline{f_1^2}$, since all higher-order contributions in (3.4) cancel after averaging /22,23/. This interesting peculiarity of the Efetov-Larkin method is lost, however, when the motion of the interface becomes relevant, i.e., when Δ/V is of the same order as the correlation time τ.

In the case $\tau \ll \Delta/V$, i.e. when the approximation $K(x,z,t) \sim K(x,z) \cdot \delta(t)$ is justified, it is easily established that

$$\overline{f_i}(x,t) \equiv 0 , \qquad \forall i. \tag{3.11}$$

Therefore $\overline{f}(x,t) = v \cdot t$ and the disorder induces a diffusion-like spreading of the front-position around its deterministic value v t only. For quenched disorder, on the other hand, (3.11) is only valid for odd i and one finds for even i:

$$\overline{f_i}(x,t) \neq 0 \quad \text{with} \quad \text{sign} \left[\overline{f_2}(x,t)\right] = - \text{sign } v.$$

Hence quenched disorder slows down the average motion of the interface, as one would expect also intuitively. The lowest-order term of the velocity correction is given by the average of (3.6), which can be written after fourier transformation as /24,25/

$$\overline{f_2}(x,t) = t \int dt' \frac{dq}{2\pi} \frac{d^{d-1}k}{(2\pi)^{d-1}} G(k,t)\ iq\ K(q,k)\ e^{iqvt'} .$$

The t'-integration can be performed to give the velocity correction $\Delta v_2 = 1/t\ \overline{f_2}$ in the form

$$\Delta v_2 = 2v \int_0^\infty \frac{df}{2\pi} \frac{d^{d-1}k}{(2\pi)^{d-1}} \frac{q^2 K(q,k)}{D^2k^4 + q^2v^2} \tag{3.12}$$

Up to the factor of order one this is equivalent to

$$\Delta v_2 \sim D^{\frac{1-d}{2}} v^{\frac{d-3}{2}} \Delta^{-\frac{d+1}{2}} K(0). \tag{3.13}$$

Consistency requires $\Delta v_2 \ll v$, i.e.

$$v \gg v_{pin} \sim [K(0)]^{\frac{2}{5-d}} D^{\frac{1-d}{5-d}} \Delta^{-\frac{1+d}{5-d}} \quad . \tag{3.14}$$

For $v \lesssim v_{pin}$ the disorder suppresses any interface motion, the interface is pinned /25/. The depinning threshold (3.14) can also be obtained by studying the height of the energy barriers between the many possible metastable configurations of a pinned interface in a random field /26/. The dependence of the actual front velocity v_{rf} in a random field on the free front velocity v without disorder as specified by (3.13) and (3.14) is schematically shown in Fig. 3. We note that (3.13) fits high-field data of non-linear charge-density wave transport well /24/. For time-dependent disorder (3.13) and (3.14) are good approximations as long as $\tau \gg \Delta/V$. If the correlation time becomes very short, $\tau \ll \Delta/V$, we find $\Delta v_2 \longrightarrow 0$ and $v_{pin} \longrightarrow 0$ as discussed above (cf. (3.11)).

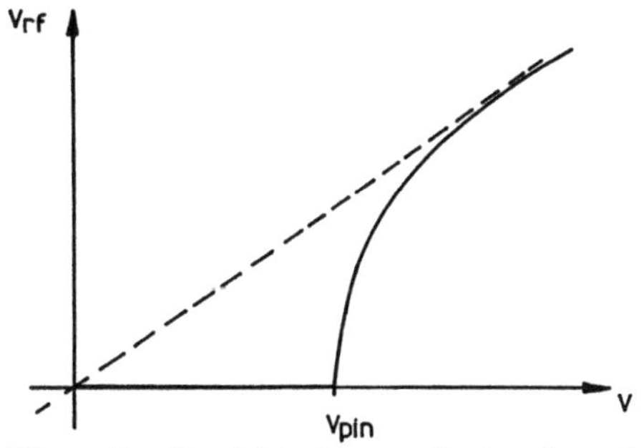

Fig. 3 Qualitative plot of actual interface velocity v_{rf} in a random field versus free interface velocity v. v_{pin} denotes the depinning threshold

New effects occur if the disorder intensity depends on the state of the system (multiplicative disorder). This can be modelled by the replacement $H(r,t) \longrightarrow g(n)\ H(r,t)$ in (1.2). A simple example is given by fluctuations $\delta k(r,t)$ of a parameter k entering the rate function $-\partial_n U(n;k)$. In this case one finds for small δk /20/:

$$H(r,t) = -\frac{\partial^2 U}{\partial n \partial k}(n;\bar{k})\ \delta k(r,t),$$

i.e.

$$g(n) = -\frac{\partial^2 U}{\partial n \partial k}(n;\bar{k}) \ .$$

For state-dependent disorder the noise intensity is different on different sides of the front, since $g(n_1) \neq g(n_3)$. This in fact gives rise to an additional term in (3.12) which does not vanish if $v \longrightarrow 0$. Therefore, an interface deterministically at rest $(U(n_1) = U(n_2))$ will move under the influence of multiplicative noise /21,27/. The motion is directed such that the part of the system being in the state with higher noise intensity decreases, so that the system tries to establish the homogeneous state with lower noise intensity. From the theory of noise-induced transitions it is well known /28-30/ that state-dependent disorder can modify the stationary probability distribution of the homogeneous state in favour of the states with lower noise intensity, i.e., with smaller values of g(n). The above results complete this picture by indicating how the preferred stationary state can be established by a system starting from an inhomogeneous initial condition /27/ (Fig. 4).

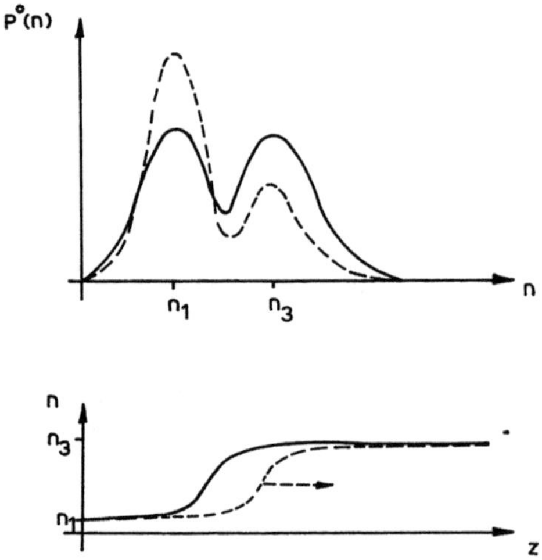

Fig. 4 Stationary probability distribution $P^0(n)$ of the homogeneous states and one-dimensional density profile of the corresponding kink solution of a bistable system with $U(n_1) = U(n_3)$. For additive noise (solid line) $P^0(n_1) = P^0(n_3)$ and the kink is on average at rest. For multiplicative noise with $g(n_1) < g(n_3)$ (dotted line) one finds $P^0(n_1) > P^0(n_3)$ and the kink moves on average to the right.

4. Summary

In this contribution several effects of disorder on moving interfaces were elucidated. A phase-separating interface of a bistable system evolves in the absence of random influences according to (2.4), which simplifies to (2.6) if only small deviations from planar occur. The second term in the r.h.s. of both equations tries to flatten an arbitrary initial configuration. Disorder, to the contrary , roughens the interfaces. This is quantitatively characterized by the statistical width as given by (3.8) and (3.10) for quenched and time-dependent randomness respectively. Moreover, quenched disorder slows down the average interface motion as given by the velocity correction (3.12). If this correction becomes of the order of the velocity itself, the interface is pinned by the randomness and is unable to move at all. The pinning threshold is given by (3.14). Velocity correction and pinning are absent if the disorder is time dependent with very short correlation time.

Finally it was sketched how state-dependent (multiplicative) noise can induce the motion of an interface deterministically at rest. This phenomenon completes the discussions of noise-induced transitions.

References

1. W. Ebeling, R. Feistel: Physik der Selbstorganisation und Evolution (Akademie-Verlag, Berlin 1982)
2. A.S. Patashinski, V.L. Pokrovski: Fluctuation Theory of Phase Transitions (Nauka, Moscow 1975 (in Russian))
3. S.M. Allen, J.W. Cahn: Acta Metall. 27, 1085 (1979)
4. A. Engel, W. Ebeling, R. Feistel, L. Schimansky-Geier: phys.stat.sol.(b), submitted

5. J.D. Gunton, M. Droz: Introduction in the Theory of Metastable and Unstable States, Lecture Notes in Physics 183 (1983)
6. M. Grant, J.D. Gunton: Phys.Rev. B28, 5496 (1983)
7. B. Ross, J.D. Litster: Phys.Rev. A15, 1246 (1977)
8. R. Landauer: Phys.Rev. A15, 2117 (1977)
9. V.V. Barelko, V.M. Beitujan, Yu.E. Volodin, Ya.B. Zeldovich: in 14
10. A.V. Gurevich, R.G. Mints: Usp.Fiz.Nauk 142, 61 (1984)
11. W. Ebeling, Yu.L. Klimontovich: Selforganization and Turbulence in Liquids (Teubner, Leipzig 1984)
12. F. Schlögl: Z.Phys. 253, 147 (1972)
13. G. Hutiray, J. Solyom (Eds.): Charge-Density-Waves in Solids, Lecture Notes in Physics 217 (1985)
14. M.T. Grekova (Ed.): Autowave Processes in Systems with Diffusion (Gorki 1981 (in Russian))
15. D.J. Wallace: Edinburgh-Preprint 1980/136
16. K. Kawasaki, T. Ohta: Progr.Theor.Phys. 67, 147 (1982); 68, 129 (1982)
17. R. Bausch, V. Dohm, H.K. Janssen, R.K. Zia: Phys.Rev.Lett. 47, 1837 (1981)
18. R. Feistel: unpublished
19. M.W. Diehl, D.M. Kroll, M. Wagner: Z.Phys. B36, 329 (1980)
20. A.S. Mikhailov, L. Schimansky-Geier, W. Ebeling: Phys.Lett. 96A, 453 (1983)
21. A. Engel: Thesis, Humboldt-Universität 1985
22. K.B. Efetov, A. Larkin: ZhETF 72, 2350 (1977)
23. A. Engel: J.de Phys.Lettre 46, 409 (1985)
24. L. Sneddon, M.C. Cross, D.S. Fisher: Phys.Rev.Lett. 49, 292 (1982)
25. M.V. Feigelman: ZhETF 85, 1851 (1983)
26. T. Nattermann: phys.stat.sol.(b) 129, 153 (1985)
27. A. Engel: Phys.Lett.A, to appear
28. R. Landauer: J.Appl.Phys. 33, 2209 (1962)
29. W. Ebeling: Phys.Lett. 68A, 430 (1978)
30. W. Horsthemke, R. Lefever: Noise-Induced Transitions Theory and Application, Springer-Series in Synergetics, Vol. 15 (Springer, Berlin Heidelberg New York 1982)

Autosolitons in Active Systems with Diffusion

B.S. Kerner and V.V. Osipov

Institute of Radio Engineering, Electronics and Automatics, Vernadsky Av. 78, SU-117 454 Moscow, USSR

The properties of a tremendously great number of physical, chemical and biological systems are described by the following system of nonlinear equations similar to the diffusion ones /1-7/:

$$\tau_i \partial X_i/\partial t = \sum_{j=1}^{N} \nabla (L_{ij} \nabla X_j) - q_i(X_1,\dots,X_i,\dots,X_n,A) \; . \qquad (1)$$

The equations (1) are of a fundamental nature, since they are chemical kinetic equations in which X_i is the temperature or concentrations of intermediate reaction products and A is the concentration of original reagents. In physics the equations (1) are based on the first three moments of the Boltzmann transport equation; they describe the distribution of the effective temperature T and that of the concentration n of hot electrons (see, for instance, /6,7/). In the latter case, X_1=T, X_2=n, and A is the external excitation power. With the parameter A having a bifurcation value A_c, a number of such systems lose their homogeneity, self-excited oscillations arise therein, and static, pulsating and/or travelling dissipative structures are formed in them /1-5,8-11/. When the lack of equilibrium is less ($A < A_c$), the homogeneous state of such a system is stable, i.e. small perturbations arising therein die away. However, when a momentary local perturbation whose amplitude exceeds a definite critical value is applied to the system, this perturbation may increase with time, the result being that the system acquires a specific localized eigenstate - a high-amplitude static, pulsating or travelling autosoliton (AS) /8-11/. The structure (Fig. 1) of ASs thus formed does not depend on the mode of the original local perturbation; it is determined only by the system parameters.

The systems characterized by these curious phenomena are referred to as active systems with diffusion. They are termed active because a positive feedback is provided for at least one of their parameters - activator ($X_i \equiv \theta$) - bringing about its self-generation and causing an instability of their homogeneous state. The activator selfgeneration process is controlled by another parameter - inhibitor $X_i \equiv \eta$ -

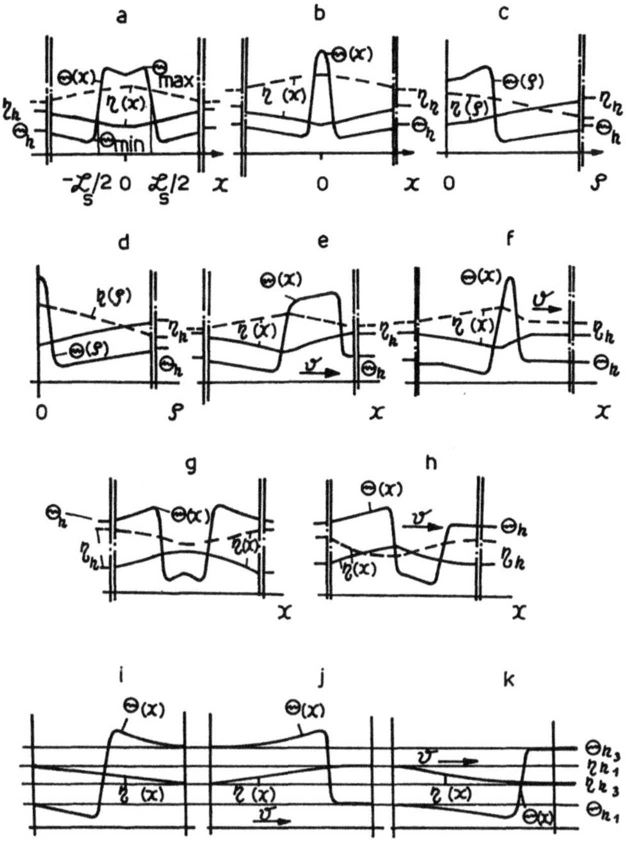

Fig. 1 The structure of the simplest AS: hot wide (a) and narrow (b) SAS; hot radially symmetrical SAS with great (c) and small radius (d); hot wide (e) and narrow (f) TAS; cold wide SAS (g) and TAS (h); SAS in a form of the complex domain wall (i) and reversible trigger waves (j) and (k)

which, as its name implies, inhibits this process. Therefore, in the simplest, though a sufficiently general case, active systems with diffusion are described by two equations /1-8/

$$\tau_\theta \frac{\partial \theta}{\partial t} = l^2 \Delta \theta - q(\theta, \eta, A) ,$$

$$\tau_\eta \frac{\partial \eta}{\partial t} = L^2 \Delta \eta - Q(\theta, \eta, A) , \qquad (2)$$

where τ_θ, τ_η and l, L are respectively the characteristic θ and η variation times and lengths. Qualitative methods of constructing, and analyzing the stability of solutions in the form of a high-amplitude AS and multiautosoliton (MAS) states in active two-component systems with diffusion have been developed by the present authors in a series of studies /7-18/. In the limiting case $L \to \infty$ /8,11,19/ the results of this general theory coincide with those of the domain theory for semiconductors with a S- or N-shaped I-V characteristic /20/, and when $L \to 0$ /11,15,16/ they coincide with the results obtained in studying autowaves (AW) in models such as the Fitz-Hugh-Nagumo (FHN) model /4,21,22/. The proposed theory relates many seemingly unrelated effects and phenomena discovered during experimental studies and computer simulations of systems differing in nature.

While studying heated electron-hole plasma (EHP) in semiconductors /6,8/, we analyzed a stratification mechanism resulting from the interdiffusion of the two components and leading, as in the case of the Turing instability /1-4/, to an aperiodic build-up of fluctuations with the wave number $k=k_0 \neq 0$. This instability in EHP causes the formation of static MASs /8,9/. The properties of hot EHP are described by two equations such as (1), in which the terms corresponding to the interdiffusion of the components X_1 and X_2 are essential /6,8,9/. To describe ASs in such systems with interdiffusion, it is also convenient to employ the concepts of the activator θ and the inhibitor η which can be presented as functions of the actual physical values X_1 and X_2 /8,9/.

1. Classification of Systems and Phenomena

For ASs of all types (Fig. 1) the inhibitor η and the activator θ change smoothly everywhere save the AS wall regions, in which the activator varies sharply over a distance of the order of l /8-10, 15/. An important feature determining the AS structure in a given system is the dependence $\eta(\theta)$ evident from the equation

$$q(\eta,\theta,A) = 0 \quad \text{at } A = \text{const.} \tag{3}$$

The above dependence will be referred to as the local relationship (LR) since this is the relationship existing between η and θ in all the regions save those of the AS walls /8-11/. In AS-containing system, the LR curve is, as a rule, N (Fig. 2a), U (Fig. 2b), Λ (Fig. 2c) or V-shaped (Fig. 2d). Accordingly, such a system will be referred to as the N-, U-, Λ- or V-System.

The values representing the homogeneous state of the system ($\theta = \theta_h$ and $\eta = \eta_h$) satisfy the equations

$$q(\eta_h,\theta_h,A) = 0 \ , \qquad Q(\theta_h,\eta_h,A) = 0 \ , \tag{4}$$

which define the point of intersection of the LR curve and the equation of state (ES) curve $\eta(\theta)$, corresponding to $Q(\theta,\eta,A) = 0$. Region I ($\theta \leqslant \theta_0$) will be termed the cold phase, region II ($\theta_0 \leqslant \theta \leqslant \theta_0'$) the heated phase and region III ($\theta \geqslant \theta_0'$) the hot phase (Fig. 2).
By analogy a system will be called cold ($A < A_0, \theta_h < \theta_0$), heated ($A_0 < A < A_0', \theta_0 < \theta_h < \theta_0'$) or hot ($A > A_0', \theta_h > \theta_0'$), depending on the phase

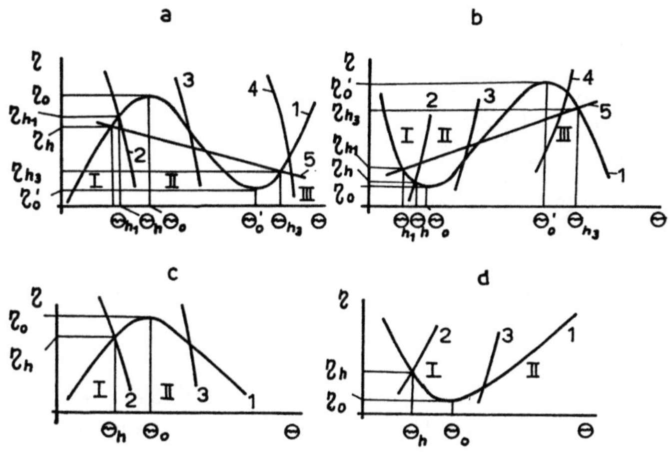

Fig. 2 Qualitative figures of the LR (curves 1) and ES (curves 2 - cold, 3 - heated, 4 - hot, 5 - trigger systems)

of its homogeneous state ($\theta=\theta_h$ and $\eta=\eta_h$, Fig. 2). The homogeneous state of a cold or hot system is stable. N- and U-systems, for which the ES graph (curve 5 in Fig. 2a and b) and the LR curve intersect at three points, are termed trigger systems. For these systems at a given A two of the three homogeneous states are stable - cold (θ_{h1}, η_{h1}) and hot (θ_{h3},η_{h3}) (Fig. 2a and b).

It follows from the general theory /8-16/ that the AS type is chiefly determined by the ratios $\varepsilon = l/L$ and $\alpha = \tau_\theta/\tau_\eta$. On this basis the systems under consideration can be divided into the following free qualitatively differing subsystems /15/:

K - systems, for which $\varepsilon \ll 1$, $\alpha > 1$;

Ω - systems with $\varepsilon \gg 1$, $\alpha \ll 1$;

KΩ - systems with $\varepsilon \ll 1$, $\alpha \ll 1$.

Systems with interdiffusion are similar to K-systems /6,8/. Existing in K-systems are only static ASs (SASs) /8-11/ or turbulence /13/. Homogeneous oscillations, pulsating ASs (PASs) and travelling ASs (TASs) cannot be produced in such systems (Fig. 3). The instability of the homogeneous state of a heated K-system results in its stratification: a sudden spontaneous formation of MAS states as periodically /8-10/ or stochastically /8,10,18/ arranged ASs filling up the system. In heated homogeneous Ω-systems self-excited homogeneous oscillations arise, only. In cold or hot systems TASs (Fig. 1e,f,h,j,k), and ASs in the form of more complex AWs can be produced /4,21,22,15/. SASs and PASs are not formed in Ω-systems (Fig. 3). In KΩ-systems SASs, TASs, PASs and more complex ASs can be generated (Fig. 3)/12,15, 16/.

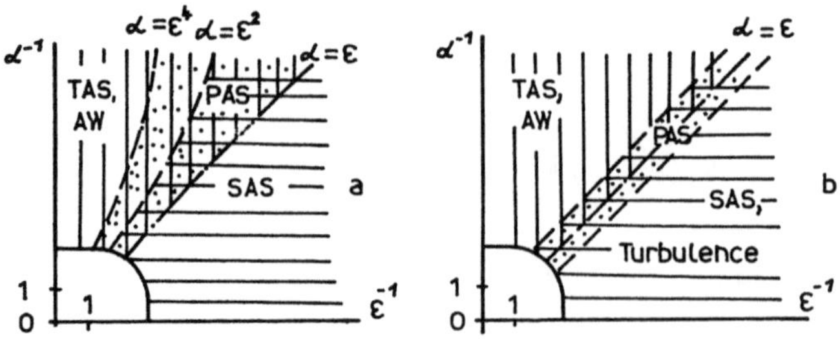

Fig. 3 Region of the existence AS in N-(U-)systems (a) and in Λ-(V-)systems (b)

The simplest SAS is a localized eigenstate (Fig. 1 a-d,g,i) in which $\eta(\vec{r})$ and $\theta(\vec{r})$ vary smoothly with the characteristic length $L \gg 1$ everywhere save the AS wall regions characterized by a sharp variation of $\theta(\vec{r})$ with the length 1 /9-11/. A PAS is an AS whose volume or shape changes periodically in time /13/. At definite moments the η and θ distribution in a PAS, i.e. its structure, resembles that of a SAS. A TAS is the simplest AW in the form of a pulse /4,21,22/ or stratum /10,11,15/ travelling at a constant velocity v (Fig. 1 e,f,h). In cold systems ($A < A_0$) hot ASs, for which θ_m within the AS kern exceeds $\theta = \theta_h$ at its periphery, can be produced (Fig. 1). On the contrary, in hot N-(U-)systems ($A > A_0'$) ASs are cold (Fig. 1 g,h). In N- and Λ-systems $\eta(\vec{r})$ varies in phase with $\theta(\vec{r})$ (solid lines, Fig. 1) everywhere save the AS wall regions; in similar AS regions of U- and V-systems the variation of $\theta(\vec{r})$ and $\eta(\vec{r})$ is out of phase (dotted lines, Fig. 1). This is in principle the only qualitative difference between AS structures in N- and U-systems. The same is true of ASs contained in Λ- and V-systems /8/. Therefore only the properties of ASs in N- and Λ-systems will be described below.

2. Autosolitons in KN- and KU-Systems

According to the terminology introduced in section 1, KN- (KU-)systems are K-systems, for which the LR curve is N- (U-)shaped (Fig. 2 a and b). Belonging to KN-systems are some axiomatic morphogenesis models /5/ and the model of gas discharge /7/ whose ASs seem to have been observed as far back as the nineteenth century by FARADAY in the form of static or moving strata /23/. An example of a KU-system is EHP heated as a result of photoexcitation or under the action of an electric field /6,9/.

In practically the entire region of existence (stability) a unidimensional SAS - solitary stratum - appears wide (Fig. 1 a,d) /9,11/: the distance between its walls $\mathcal{L}_s \gg 1$. Allowing for the SAS symmetry relative to its centre (x=0, Fig. 1a), the θ and η distributions in a hot SAS can be written to within $\varepsilon \ll 1$ as

$$\theta(x) = \theta_{sh}(x) + \begin{cases} \theta_{III}-\theta_{s3} \\ \theta_{I}-\theta_{s1} \end{cases} \tag{5}$$

$$\eta(x) = \begin{cases} \eta_{III}, 0 \leqslant x \leqslant \mathcal{L}_s/2 \\ \eta_I, \mathcal{L}_s/2 \leqslant x < \infty \end{cases}$$

Here, $\theta_{sh}(x)$ is a sharply varying distribution describing a SAS wall and corresponding to the separatrix for the equation

$$l^2 d^2\theta/dx^2 = q(\theta, \eta_{sh}, A) \qquad \eta_{sh} = \text{const.} \tag{6}$$

and $\eta_{I,III}(x)$, $\theta_{I,III}(x)$ are smoothly varying distributions which describe $\eta(x)$ and $\theta(x)$ everywhere save the AS wall regions and are consistent with the equations

$$L^2 \frac{d^2\eta_j}{dx^2} = Q_j \equiv Q(\theta_j(\eta), \eta; A) ,$$

$$q(\theta_j, \eta, A) = 0 , \qquad j = I, III \tag{7}$$

whose solutions satisfy the limiting conditions $\eta_I(\infty) = \eta_h$,

$$\eta_I(\mathcal{L}_s/2) = \eta_{III}(\mathcal{L}_s/2) = \eta_s \qquad d\eta_{III}/dx\Big|_{x=0} = 0 .$$

The subscripts I and III indicate that the relationship between θ and η is described by the Ist and IIIrd branch of the LR curve of the unique dependence $\theta(\eta)$ (Fig. 2a). In (6) the value of η in a wall of a wide AS ($\eta_{sh} = \eta_s$) can be determined from equations

$$\int_{\theta_{s1}}^{\theta_{s3}} q(\theta, \eta_s, A) d\theta = 0 , \qquad q(\theta_{si}, \eta_s, A) = 0, \qquad i = 1,2,3 \tag{8}$$

which also determine the minimum ($\theta_{min} = \theta_{s1}$) and maximum ($\theta_{max} = \theta_{s3}$) values for a SAS (Fig. 1a) and the θ values at the point $x = \mathcal{L}_s/2$: $\theta_{sh}(\mathcal{L}_s/2) = \theta_{s2}$. Integrating (7) with allowance for the continuity of the inhibitor flow through an AS wall (for $x = \mathcal{L}_s/2$, Fig. 1a) we obtain the following equations for determining $\eta(0) = \eta_m$ and $\theta(0) = \theta_m$ at the SAS centre and its width $\mathcal{L}_s$:

$$q(\eta_m, \theta_m, A) = 0, \ \int_{\eta_s}^{\eta_h} Q_I d\eta + \int_{\eta_m}^{\eta_s} Q_{III} d\eta = 0, \ \mathcal{L}_s = \sqrt{2} L \int_{\eta_m}^{\eta_s} \Big[\int_{\eta_m}^{\eta_s} Q_{III} d\eta \Big]^{-1/2} d\eta . \tag{9}$$

Thus, the principle SAS parameters (θ_{max}, θ_{min}, θ_m, η_m, η_s and $\mathcal{L}_s$) can be determined to within ε from simple algebraic equations (8) and (9). To determine the SAS stability region in KN-systems, it will suffice to find the η_{sh}-A dependence /8-11/ - the bifurcation characteristic of the system (Fig. 4). Over a wide range of variation

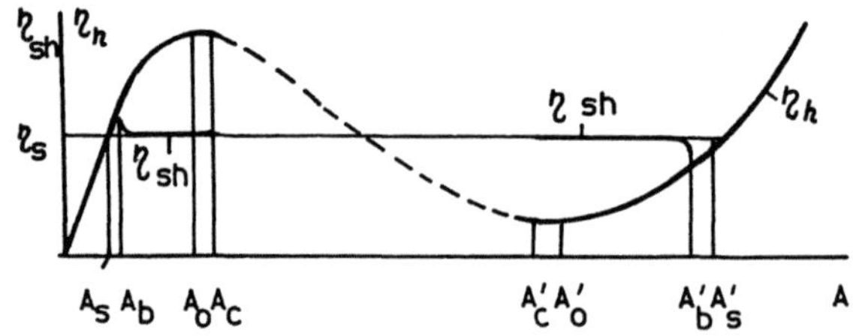

Fig. 4 Possible bifurcation characteristic of the N-system

of A this dependence coincide with the dependence $\eta_s(A)$ which follows from (8). By analyzing (9) it can be concluded that while a system cools (A decreases) the hot SAS width is reduced, and at $A = A_b$ when the SAS width $\mathcal{L}_s = \mathcal{L}_b \sim 1 \cdot \ln \varepsilon^{-1}$ in the vicinity of the point, at which $d\eta_{sh}/dA = \infty$ (Fig. 4), the SAS disappears suddenly /9-11,18/.

As A increases the SAS becomes wider and at $A > A_c$ a MAS is formed spontaneously. A cold SAS (Fig. 1g) existing in a hot system at $A'_c \leq A \leq A'_b$ (Fig. 4) behaves similarly. For systems where $q(\theta,\eta)$ and, consequently, η_s are independent of A (this is, for instance, characteristic of the gas discharge /7/), the bifurcation characteristic is of the form shown in Fig. 4. For such systems $A_b = A_s$ and $A'_b = A'_s$ to within ε (A_s and A'_s are the A values that are true for $\eta_h = \eta_s$ (Fig. 4)).

In two- and three-dimensional systems a unidimensional SAS (Fig. 1a) exists within a narrower range of variation of A than in unidimensional systems. At the limits of this range, the SAS becomes unstable to the goffering of its walls /10,16/. The width $\mathcal{L}_s$ of a stable SAS is within $(\mathcal{L}_{b1}, \mathcal{L}_{c1})$ /16/

$$\mathcal{L}_b < \mathcal{L}_{b1} \sim 1 \ln \varepsilon^{-3/2} < \mathcal{L}_s < \mathcal{L}_{c1} \sim (1L)^{1/2} . \qquad (10)$$

In a three-dimensional system a radially symmetrical SAS may develop in the form of a drop (Fig. 1c) or hollow globe, and in a two-dimensional system it can take the form of a spot (Fig. 1c) or ring /10/. With radius of such SASs $\varrho_0 \gg 1$, the distributions $\eta(\varrho)$ and $\theta(\varepsilon)$ are described to within ε by equations (5),(6),(8) and (7) if d^2/dx^2 in equation (7) is substituted by $\varrho^{-(1+s)}(d/d\varrho)(\varrho^{+(1+s)}d/d\varrho$ (for a drop s=1, for a spot S=0). Even at $A_b < A < A_c$ (Fig. 4) a drop may become unstable to radially unsymmetrical fluctuations goffering its surface and breaking the drop /10,16/.

At a given A within the $A_b < A < A'_b$ range, a system incorporates a multitude of various stable MASs in the form of symmetrical and unsymmetrical periodically and stochastically arranged SASs /8,10,18/. In two- and three-dimensional systems MASs can be produced in the form of a rather complex and contrasting pattern. The presence of complex MASs and their predicted form /8,10/ were confirmed by computer simulations /24/ and analytically provided in /18/. When a homogeneous state is stratified ($A=A_c$), the critical fluctuation period $\mathcal{L}_p = 2\pi k_0^{-1}$ /2,3/. In unidimensional KN- and KU-systems with $\mathcal{L} \gg L$, MASs characterized by this period are stable /9-11/; therefore, the stratification results, as a rule, in the formation of $N = \mathcal{L}/\mathcal{L}_p$ periodically arranged SAS strata. As A drops resulting in the period-doubling bifurcations (pumpover effect) the number of strata will decrease stepwise /8,9/ so that at A close to A_b a single hot SAS is spontaneously produced in the system (Fig. 1a). On the contrary, as A rises, the hot strata are expanded and transformed into cold ones; at $A \to A_b$ the pumpover causes a stepwise decrease in the number of these cold strata, resulting in the formation of a solitary cold SAS in the system (Fig. 1g). The complex MAS evolution theory results /8-11/ including those on two- and three-dimensional MASs are reviewed in /25/.

3. Autosolitons in ΩN- and ΩU-Systems

Belonging to Ω N-systems are models such as the FHN /4,21,22/ one and that of semiconductor plasma /15,26/. An example of an Ω U-system is the oregonator model /22/.

In hot and cold systems, ASs can be generated as AWs, of which the simplest ones are TASs, single-arm and multi-arm reverberation units (RU), scrolls and rings /22/. As distinct from wide SASs (section 2), in two- and three-dimensional systems a wide TAS is stable to goffering /15,16/ which accounts for the existence of RUs and other complex AWs.

4. Autosolitons in KΩN- and KΩU-Systems

The conditions $\varepsilon = l/L \ll 1$ and $\alpha = \tau_\theta/\tau_\eta \ll 1$ determining KΩ-systems are also true for a great number of physical, chemical and biological systems, since for the majority of them $l = (D_\theta \tau_\theta)^{1/2}$, $L = (D_\eta \tau_\eta)^{1/2}$ and $D_\eta \simeq D_\theta$ /2,4,13,16,27/.

For both KΩ- and K-systems $\varepsilon \ll 1$; therefore, for KΩ-systems the solutions of the steady-state equations (2) resemble those of the same equations for K-systems. The properties of SASs in KΩ-systems with $\varepsilon \ll \alpha \ll 1$ and in K-systems coincide fully /15/. Owing to a small value of α and the resulting possibility of θ and η frequency isolation, SASs existing in KΩ-systems with $\alpha \lesssim \varepsilon \ll 1$ appear less stable than those incorporated in K-systems. They become unstable to pulsations, i.e. to augmented fluctuations localized in the AS walls and oscillating at the separated frequency /13/. The simplest PASs are states in the form of a "breathing" or "twisting" stratum wall, solitary pulsating spot or globe /13,16/. At its periphery, a PAS can generate diverging undamped AWs /13/, i.e. it can play part of a leading centre /4,22/. Apart from SASs and PASs, TASs moving at the velocity $v \sim 1/\tau_\theta$ and other more complex AWs can be produced in KΩ-systems, for which $\varepsilon^2 < \alpha \lesssim \varepsilon \ll 1$ /15/. In such systems a change in A may cause a spontaneous transformation of SASs to PASs having the frequency $\omega \sim (\alpha\varepsilon)^{1/2} \tau_\theta^{-1}$ or to TASs moving at $v \sim (\alpha/\varepsilon)^{1/2} 1/\tau_\theta$ /15, 16/. In systems with $\varepsilon^4 < \alpha < \varepsilon^2 \ln \varepsilon^{-2} \ll 1$ all SAS are unstable, and in such systems PASs and ASs can be produced in the form of AWs (Fig. 3a). However, in KΩ-systems at $\alpha < \varepsilon^4$ only AWs including TASs whose velocity v is such that $\alpha^{1/2} 1/\tau_\theta < v \lesssim 1/\tau_\theta$ can be generated (like in Ω-systems) (section 3) /15/. The range of A values within which TASs exist is maximum at $\alpha < \varepsilon^4$, and it is the more narrow, the higher ratio α/ε /15/.

5. Autosolitons in Λ- and V-Systems

Belonging to V-systems are the Gierer-Meinhardt morphogenesis model /3,5/ and the EHP model for some semiconductors. Brusselator /2/, heated EHP /16/ and gas discharge /7/ models are examples of Λ-systems. In cold unidimensional systems only narrow hot ASs are produced - a SAS in K-systems (Fig. 1b) /8/, a TAS in Ω-systems (Fig. 1f) and a SAS, TAS and PAS resembling in narrow varying-amplitude SAS in KΩ-systems (Fig. 3b) /13,15/. In two- and three-dimensional systems unidimensional, ASs are unstable to fluctuations, tending to break them into parts, the shape of which resembles a spot or a globe having the radius $\rho_0 \sim 1$ (Fig. 1d). Cylindrical, spherical and spiral AWs are unstable, too. Only ASs, all dimensions of which are small (of the order of 1) can be stable /10,15/. A unique feature of Λ - and V-systems is that the less ε, the lower level of nonequilibrium A, at which AS can be formed therein, and the higher AS amplitude. In other words, a short-term high-amplitude local perturbation can produce a stable AS in almost equilibrated physical systems with $\varepsilon \ll 1$ /17/.

The stratification of the homogeneous state of a unidimensional or three-dimensional heated K-system causes respectively a sudden formation of narrow strata /8/ or small-radius drops /10/ spaced periodically or stochastically at $\mathcal{L}_p \gg 2\pi k_0^{-1}$; those ASs that are spaced at $2\pi k_0^{-1}$ are, as a rule, unstable to the pumpover effect /8/. As A rises the number of ASs in the system will increase in steps owing to the local breakdown across them /8,12/. In some heated K-systems all steady states are unstable and irregularly oscillating MASs in the form of spontaneously apearing and disappearing MASs - a kind of turbulence - are produced /13/.

References

1. W. Ebeling: Strukturbildung bei irreversiblen Prozessen (Teubner, Leipzig 1976)
2. G. Nicolis and I. Prigogine: Self-Organization in Nonequilibrium Systems (Wiley, New York 1977)
3. H. Haken: Synergetics (Springer, Berlin Heidelberg New York 1978)
4. V.A. Vasilev, Yu.M. Romanovski, V.G. Yakhno: Sov.Phys.Usp. 22, 615 (1979)
5. Yu.M. Romanovski, N.V. Stepanova, D.S. Chernavski: Matematicheskaya biofisika (Nauka, Moscow 1984)
6. B.S. Kerner, V.V. Osipov: Sov.Phys.JETP 44, 807 (1976); Sov.Phys. Sem. 13, 523 (1979): Sov.Phys.Solid St. 21, 1348 (1979)
7. B.S. Kerner, V.V. Osipov: Sov.Phys. - Dokl. 26, 420 (1981); Sov.Radio Eng & Electron.Phys. 27, 122 (1982); 28, 119 (1983)
8. B.S. Kerner, V.V. Osipov: Sov.Phys.JETP 77, 874 (1978)
9. B.S. Kerner, V.V. Osipov: Sov.Phys.Semicond. 13, 424 (1979)
10. B.S. Kerner, V.V. Osipov: Sov.Phys.JETP 52, 1122 (1980)
11. B.S. Kerner, V.V. Osipov: Sov.Microelectronics (C/b New York) 10 (1981)
12. B.S. Kerner, V.V. Osipov: Sov.Phys. - Dokl. 27, 484 (1982)
13. B.S. Kerner, V.V. Osipov: Sov.Phys.JETP 56, 1275 (1982); Sov.Phys. - Dokl. 28, 485 (1983)
14. B.S. Kerner, V.V. Osipov: Biofisika 27, 137 (1982)
15. B.S. Kerner, V.V. Osipov: Sov.Microelectronics (C/b New York) 12, (1983)
16. B.S. Kerner, V.V. Osipov: Zh.Eksp.Teor.Fiz. 89, 2 (1985) (Sov.Phys. JETP 1985)
17. B.S. Kerner, V.V. Osipov: Pisma Zh.Eksp.Teor.Fiz. 41, 386 (1985) (JETP Lett. 1985)
18. B.S. Kerner, V.V. Osipov: DAN SSSR 277, 1114 (1984) (Sov.Phys. Dokl. 1984); Sov.Microelectronics (C/b New York) 13 (1984)
19. B.S. Kerner, V.V. Osipov: Sov.Microelectronics 6, 256 (1977)
20. A.F. Volkov, Sh.M. Kogan: Sov.Phys.Usp. 11, 881 (1968)
21. L.A. Ostrovski, V.G. Yakhno: Biofisika 20, 489 (1975); R.G. Casten, H. Cohen, P.A. Lagerstron: Quart.Appl.Math. 32, 365 (1975)
22. L.S. Polak, A.S. Michailov: Samoorganisatsia v neravnovesnih fisiko-cimicheskih sistmah (Nauka, Moscow 1983)
23. L. Pekared: Usp.Fiz.Nauk. 94, 463 (1968)
24. Yu.I. Balkarei, M.G. Evtikhov, M.I. Elinson: Sov.Microelectr. 8, Nr. 6 (1982); 9, Nr. 2 (1980); 10, Nr. 1 (1981); 11, Nr. 1 (1982); 12, Nr. 1,2 (1983)
25. B.S. Kerner, V.V. Osipov: Sov.Microelectronics 14, Nr. 5 (1985)
26. Yu.A. Pjanov, Yu.I. Balkarei, L.L. Golik, M.I. Elinson: Fiz.Tekh.Poluprov. 17, Nr. 2 (1983)
27. Yu.I. Balkarei, M.G. Nikulin: Sov.Phys.Semicond. 10, 1216 (1976)

Repressing of Chemical Waves by Photochemical Inhibitor Releasing

L. Kuhnert, L. Pohlmann, H.-J. Krug, and G. Wessler

Zentralinstitut für Physikalische Chemie der Akademie der Wissenschaften der DDR, Rudower Chaussee 5, DDR-1199 Berlin, GDR

The Belousov-Zhabotinsky (BZ) reaction exhibits a very broad range of complex dynamic behaviour. The occurrence of chemical waves (also called trigger waves or autowaves) in a thin layer of unstirred BZ reagent is a well-known phenomenon. There are a lot of experimental and theoretical investigations dealing with this kind of waves (see e.g., /1,2/). It is well known that the necessary condition for the formation of travelling fronts, pulses or wave trains is an excitable (bistable or excitable monostable) or an oscillatory (limit cycle) behaviour of the corresponding homogeneous reaction system. For the sake of simplicity we restrict here our considerations to the propagation of travelling reaction fronts in the incomplete BZ system, i.e. in the BZ reagent without organic compounds like malonic acid. These autocatalytic reaction systems show similar characteristic properties, like, e.g. constant front velocity and nucleation phenomena, as the complete BZ system.

In principle, there is still some gap between theoretical descriptions of wave fronts and nucleation phenomena and their experimental observations in BZ reactions. That means, the profound investigations of the BZ reaction mechanism (compare /1,3/, e.g.) predict a reaction system which is either excitable monostable or oscilatory in dependence on system parameters. However, this system has no bistable properties. The same statement is also valid for one- or two-component systems obtained from BZ systems by mathematical reducing procedures (compare /4/ in the case of the Oregonator).

On the other hand, in the theory of travelling reaction fronts two different kinds of front solutions exist. The first one is exhibited in a system with one stable and one unstable stationary state of the corresponding homogeneous system. These fronts have a continuum of front shapes with a continuum of front velocities bounded from below /5/. Also, it seems to be that nucleation of a new (globally stable) phase in this system occurs with maximal probability because the so-called supersaturation defined by EBELING and SCHIMANSKY-GEIER /9/ is here always maximal. The second kind of front solution occurs in systems with bistable behaviour /6/. These fronts have a unique shape and only one well-defined velocity. Nucleation probability here depends on the value of supersaturation which may change from zero to maximal values /7,8,9/. Also, in the case of weak supersaturation, nucleation processes are very seldom. In fact, mainly these properties corresponding to bistable systems are observed in the BZ reaction in a thin unstirred layer of reagent.

However, new experimental investigations of ruthenium catalyzed BZ systems /10/ on the one hand and of oxygen influence on BZ reaction /11/ on the other hand show us a way to close this gap of understanding BZ reactions in reaction-diffusion (RD) systems.

First of all, it should be noted that bistable behaviour was experimentally observed and theoretically described also in BZ reactions, but this behaviour is possible in stirred open systems with permanent inflow of reagent only /12,13/. Secondly, because the autocatalytic reaction step in BZ reaction is essentially to first order we are not able to obtain (by mathematical reducing procedures) a realistic one-component model of BZ reaction with bistable properties like, e.g. the Schlögl model.

However, the above-mentioned experimental investigations show us a possibility to find out a realistic two-component system with bistable properties. The crux is that both outer influences on BZ reaction, the photochemical excitation of Ru^{2+} as well as the oxidation of bromomalonic acid by oxygen lead to an additional source of the inhibitor bromide. Therefore, to the reaction equations describing BZ reaction (see, e.g. /14/) in the Oregonator model

$$BrO_3^- + Br^- + 2\,H^+ \xrightarrow{K_1} HBrO_2 + HOBr \qquad (1)$$

$$Br^- + HBrO_2 + H^+ \xrightarrow{K_2} 2\,HOBr \qquad (2)$$

$$BrO_3^- + HBrO_2 + 3\,H^+ + 2\,Ru^{2+} \xrightarrow{K_3} 2\,Ru^{3+} + 2\,HBrO_2 + H_2O \qquad (3)$$

$$2\,HBrO_2 \xrightarrow{K_4} HOBr + BrO_3^- + H^+ \qquad (4)$$

$$4\,Ru^{3+} + BrCH(COOH)_2 + 2\,H_2O \xrightarrow{K_5} 4\,Ru^{2+} + HCOOH + 2\,CO_2 + 5\,H^+ + Br^- \qquad (5)$$

$$Ru^{2+} = R(bpy)_3^{2+}$$

we must add a bromide source caused by excitation of Ru^{2+} in our case:

$$Ru^{2+} \xrightarrow{h\nu} {}^*Ru^{2+} \qquad (6)$$

$$6\,{}^*Ru^{2+} + BrO_3^- + 6\,H^+ \longrightarrow 6\,Ru^{3+} + 3\,H_2O + Br^- \qquad (7)$$

By absorption of light a dramatic change in the redox properties of Ru^{2+} takes place.

$$Ru^{3+} + e^- \longrightarrow Ru^{2+} \qquad E_0 = 1.26\ V$$

$$Ru^{3+} + e^- \longrightarrow Ru^{2+} \qquad E_0 = -\,0.84\ V \quad /17/$$

In contrast to the ground state it is evident that the excited molecule is expected to be a very strong reducing agent.

$^*Ru^{2+}$ in the excited state is able to reduce bromate to the inhibitor bromide directly via (7).

For the sake of simplicity, we introduce in our model two restrictions. Firstly, we investigate only the incomplete BZ system, i.e. without bromomalonic acid. Secondly, we treat the bromide producing reactions (6,7) as a quasi-constant bromide source Φ which is proportional to the intensity of light. It should be noted out here that this assumption is only valid if catalyst Ru^{2+} is always in excess. The experiments (Fig. 3) were always carried out under conditions of total absorption of the incident light by Ru^{2+}.

Accordingly, the RD equations are the following:

$$\partial X/\partial t = \partial^2 X/\partial z^2 + X(1-X) + (L-X)Y \ ; \quad X \equiv [HBrO_2] \qquad (8)$$

$$\partial Y/\partial t = \partial^2 Y/\partial z^2 + \varphi - Y(bX+M) \ ; \quad Y \equiv [Br^-] \qquad (9)$$

with the scaling units

$$X_0 = \frac{K_3 A}{2K_4}\ ;\ Y_0 = \frac{K_3 A}{K_2}\ ;\ z_0 = \left(\frac{D}{K_3 A}\right)^{1/2};\ t_0 = (K_3 A)^{-1}\ ;\ A \equiv [BrO_3^-]$$

the parameters

$$M = \frac{K_1}{K_3} \ ; \ b = \frac{K_2}{2K_4} \ ; \ L = \frac{M}{b} \ ; \ \varphi = \frac{K_2 \Phi}{(K_3 A)^2}$$

and with equal diffusion coefficients assumed.

Let us now apply the usual techniques of bifurcation theory (compare e.g. /15/) to the corresponding homogeneous system with φ as bifurcation parameter. If we drop the small parameter L ($L \sim 10^{-3}$) we obtain a typical bifurcation diagram with a transcritical bifurcation point at $\varphi = M$, $X = 0$. In Fig. 1 are shown the two solution branches, the trivial one $X \equiv 0$ which bifurcates at $\varphi = M$.

Further, after reintroduction of L, the bifurcation diagram changes qualitatively, i.e. the small perturbation disturbs the bifurcation and we obtain an imperfect bifurcation (see /16/). Now there are two isolated solution branches (see Fig. 2). However, only the upper one is relevant to us because the other leads to $X < 0$ only when $\varphi > 0$.

Therefore, at $0 < \varphi < \varphi_0$ (because of the smallness of L, φ_0 is close to M and φ_M^* to φ_M^0) only one, naturally stable, stationary state (the oxidized phase) exists. At $\varphi_0 < \varphi < \varphi_M^*$, the system has two stable and one unstable stationary states, as stability analysis shows. If $\varphi > \varphi_M^*$, two stationary states disappear and only the lower one (the reduced phase) remains. In analogy to one-component bistable systems, we now conclude that there exists a value $\varphi_0 < \varphi_K < \varphi_M$ so that with $\varphi_0 < \varphi < \varphi_K$ the upper solution is globally stable and the front moves in the direction of the lower stable state. In the opposite case, $\varphi_K < \varphi < \varphi_M$, the lower state is globally stable. Finally, at $\varphi = \varphi_K$ the front velocity vanishes and coexistence between the two stable states appears in the RD system.

Consequently, the bromide source φ is a new control parameter in a RD system containing the photosensitive BZ reaction with Ru as catalyst. With changes of the light intensity one can control the appearance or disappearance, respectively, of stable travelling fronts in dependence on φ. Also, φ controls whether the front moves in the direction of the reduced or the oxidized phase. In the marginal case $\varphi = \varphi_0$, both the front velocity and the supersaturation are maximal, whereas at $\varphi = \varphi_K$ both vanish. If $\varphi > \varphi_M^*$, only the reduced phase may exist and wave fronts are completely repressed.

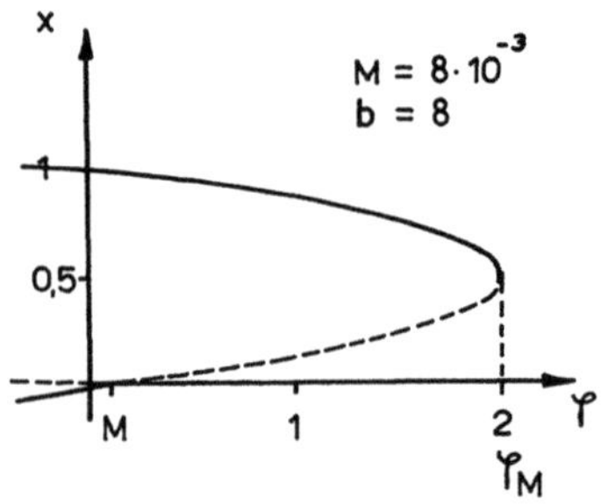

Fig. 1 Bifurcation diagram with L = 0
(—— stable stationary states, ---- unstable states)
$\varphi_M = 2.0005$

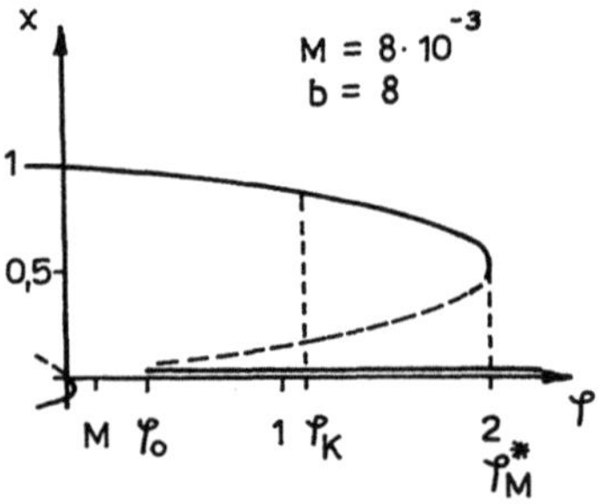

Fig. 2 Diagram of imperfect bifurcation with $L = 10^{-3}$
$\varphi_M^* = 2.0005$; $\varphi_0 = 0.0465$
(—— stable states, ---- unstable states)

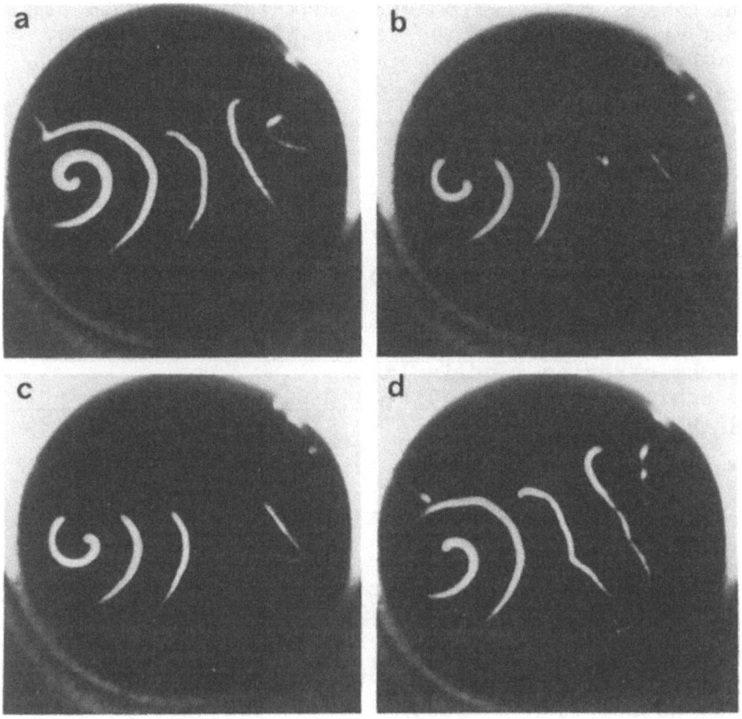

Fig. 3 Repressing of chemical waves by photochemical action
Initial concentrations were: $NaBrO_3$=0.186; H_2SO_4=0.003; Malonic acid = 0.027, Bromomalonic acid =0.063, $Ru(bpy)_3SO_4$ =0.004 M. Thermostated petridish 25°C, diameter: 7 cm, depth of reaction layer 1.3 mm. Illumination from below by a Argon-laser beam λ = 445 nm.
3a: Dark reaction
3b: Illumination of the upper half 80 sec
3c: Illumination of the upper half 120 sec
3d: Relaxation of the whole medium 120 sec

Calculations with realistic parameters show that in daylight φ is of order of magnitude of φ_0 therefore, in dependence of concentrations of reactants the system may be in the bistable state or in the monostable one. Under these conditions the distributed Ru-system has properties like ferroin /18/.

Accordingly, $\varphi \ll \varphi_K$ and only with very high light intensity one may reach φ_M^* and completely repress the occurrence of wave fronts.

An experimental example is shown in Figs. 3a-d. Firstly (3a) the dark reaction is demonstrated. In 3b and c the upper half of the layer was exposed, in the illuminated region all waves were destroyed. By photochemical inhibitor releasing via reaction (6) and (7) we have realized a value $\varphi > \varphi_M$ in the exposed area.

In the upper half the conditions were not altered. The wave trains move with constant velocity (~1,2 mm/min). The core of the spiral in the dark region was not influenced, but the spiral arm was cut through the light beam. After a relaxation time of 120 sec (3d) the inhibitor concentration has dropped by reaction (1), and we have reached again a value $\varphi_0 < \varphi < \varphi_K$ in the whole medium and wave trains can move in the formerly exposed upper half.

Also we have made observations which show that the mechanism of wave repressing goes via stopping of the front, if we have reached $\varphi = \varphi_K$ by light action.

References

1. A.M. Zhabotinsky: Konzentrazionnye avto kolebanya (Moscow 1974)
2. L. Kuhnert, H.-J. Krug, L. Pohlmann: J.Phys.Chem. 89, 2022 (1985)
3. R.J. Field, E. Körös, R.M. Noyes: J.Am.Chem.Soc. 94, 8649 (1972)
4. J. Tyson: J.Phys.Chem. 86, 3006 (1982)
5. A. Kolmogorov, I. Petrowsky, N. Piskunov: Univ.Moscow Bull. Math. 1, 1 (1937)
6. F. Schlögl: Z.Physik 253, 147 (1972)
7. W. Ebeling, R. Feistel: Physik der Selbstorganisation und Evolution (Akademie-Verlag, Berlin 1982)
8. H. Malchow, L. Schimansky-Geier: Noise and Diffusion in Bistable Nonequilibrium Systems, Leipzig 1985 (to appear)
9. W. Ebeling, L. Schimansky-Geier: Proc. 6th Int.Conf.Thermodyn. p. 65, Merseburg 1980
10. L. Kuhnert: Naturwiss. (in press)
11. S. Barkin, M. Bixon, R.M. Noyes, K. Bar-Eli: Int.J.Chem.Kin. 10, 619 (1978)
12. W. Geiseler, H. Föllner: Biophys. Chem. 6, 107 (1977)
13. P. DeKepper, J. Boissonade: J.Chem.Phys. 75, 189 (1981)
14. J.D. Murray: Lectures on Nonlinear-Differential-Equation Models in Biology (Oxford 1977)
15. G. Iooss, D. Joseph: Elementary Stability and Bifurcation Theory (Springer, Berlin Heidelberg New York 1980)
16. B. Matkowsky, E. Reiss: SIAM J.Appl.Math. 33, 230 (1977)
17. N. Sutin, C. Creutz: Adv.in Chemistry series No. 168, Inorganic and Organometallic Photochemistry 1 (1978)
18. L. Kuhnert, H.-J. Krug: J.Phys.Chem. (submitted)

Dynamic Patterns in Interacting Chemical Cells and Effects of External Periodic Forcing

M. Marek, M. Dolnik, and I. Schreiber

Department of Chemical Engineering, Prague Institute of Chemical Technology, CS-166 28 Prague, Czechoslovakia

1. Introduction

Chemical reactions with nonlinear kinetics serve as model systems for the study of nonlinear dynamics /1/. Periodic, quasiperiodic and chaotic oscillations are observed in a flow-through well-stirred reaction cell /2/. A large number of dynamic systems are modelled as a set of oscillating units with mutual interaction. The interaction is one-directional in a special case of periodically forced systems. Well-stirred flow-through reaction cells with an oscillating chemical reaction and mutual mass exchange and/or external concentration periodic forcing serve as a convenient model system /3-6/ with a number of analogies, for example, to biological cell structures. The Belousov-Zhabotinsky reaction /7/ is most often used as a source of oscillations in a single reaction cell.

In this paper we summarize experimental results obtained with two experimental systems: a system of two reaction cells with mutual mass exchange and a system of two coupled reaction cells with one cell forced by external periodic concentration pulse forcing.

2. Two Coupled Chemical Oscillators

The experimental system of two identical thermostated well-stirred reaction cells was described elsewhere /8/. The frequency of oscillations in uncoupled cells was controlled by varying the intensity of stirring. The relative intensity of interaction D was controlled by varying the area of apertures between the cells. The concentration oscillations were followed by measuring redox-potential with Pt wire electrodes, cf. Fig. 1.

The oscillations under given reaction conditions are relaxational and the dynamic regime can be described by means of the sequence of "periods", i.e., the intervals between the subsequent jumps of the potential (cf. lower part of Fig. 1). Times of the potential jumps are shown below as vertical bars; the longer bars correspond to the faster and the shorter bars to the slower cell.

The uncoupled cells oscillate with frequencies f_F ("faster" cell), f_S ("slower" cell) ($f_F > f_S$). The average frequencies over the entire experiment are defined as

$$f_F = n_F/T , \quad f_S = n_S/T .$$

Here n_F(n_S) denotes the number of oscillations of the faster (slower) oscillator over time T.

The synchronization ratio S is defined as

$$S = \frac{f_F}{f_S} = \frac{n_F}{f_S} .$$

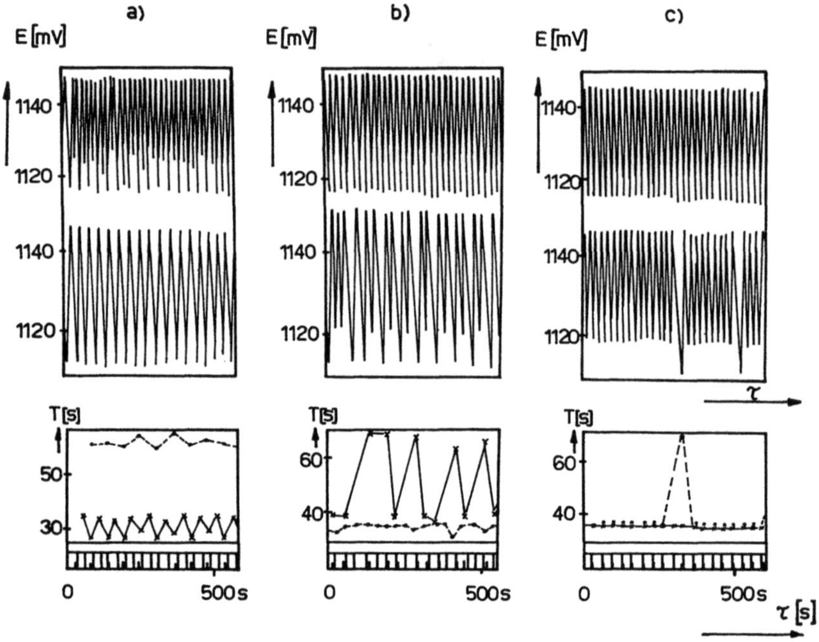

Fig. 1 Recordings of redox-potential and sequences of "periods". Initial ratio of frequencies f_F/f_S = 1.40
a) D = 12, periodic pattern with S = 2.00
b) D = 13, aperiodic pattern with S = 1.49
c) D = 14, aperiodic pattern with S = 1.06.

If S can be written as a ratio of small integers, then we can expect periodic, synchronized regime. However, the synchronization ratio S describes only global behaviour of the system and instanteneous synchronization or desynchronization of the oscillator is not detected. Hence, making use of the relaxational character of the oscillations, we divide the entire length of the experiment T into intervals T_k in such a way that the neighbouring intervals T_k, t_{k+1} are separated by the time of a synchronous potential jump in both reactors. Thus the phase shift between the oscillators is close to zero (less than the level of experimental noise) at this jump. On each interval T_k we then define "local" synchronization ratio S_k in a similar way to S. These ratios can be grouped according to their values, so that the i-th group contains S_k s with a common value s_i. If the i-th group contains n_i elements, then the relative frequency of their occurrence $I_i=n_i/\sum_i n_i$. The values of S and I_i s for the series of experiments with various values of parameters f_F/f_S and D are given in Table 1. The relative frequency of occurrence clearly differentiates between periodic and aperiodic regimes and points in several cases to the existence of intermittency, i.e. longer series of regular oscillations are disturbed by another rhythm, cf. Fig. 1c and the corresponding data in Table 1.

Table 1

f_F/f_S	D	8		12		16		20	
	S	1.39		1.20		1.07		1.00	
	i	s_i	I_i	s_i	I_i	s_i	I_i	s_i	I_i
1.27	1	3/2	0.11	5/4	0.34	5/4	0.30	1/1	1.00
	2	10/7	0.10	6/5	0.46	1/1	0.70		
	3	7/5	0.38	7/6	0.14				
	4	4/3	0.41	1/1	0.06				
	D	8		10		12		14	
	S	1.50		1.57		1.47		1.40	
	i	s_i	I_i	s_i	I_i	s_i	I_i	s_i	I_i
1.34	1	3/2	1.00	5/3	0.23	3/2	0.77	2/1	0.04
	2			8/5	0.24	4/3	0.23	5/3	0.03
	3			11/7	0.10			3/2	0.63
	4			3/2	0.46			4/3	0.16
	5							1/1	0.14
	D	10		12		13		14	
	S	2.00		2.00		1.49		1.06	
	i	s_i	I_i	s_i	I_i	s_i	I_i	s_i	I_i
1.40	1	2/1	1.00	2/1	1.00	2/1	0.66	2/1	0.11
	2					1/1	0.34	1/1	0.89

3. Coupled Cells with External Pulse Forcing

Dynamic behaviour of a single cell with pulse periodic forcing was discussed in an earlier paper /6/. This system contains a one-directional coupling between natural oscillations and external pulse forcing and we shall call it a "nonautonomous oscillator". The dynamic regimes in the nonautonomous oscillator are controlled by an amplitude and relative frequency τ_F of the external periodic perturbation ($\tau_F = T_F/T_B$, where T_F is the forcing period and T_B is the period of the unperturbed oscillator).

When another cell (an autonomous oscillator with the period T_B) is coupled with the nonautonomous oscillator, complex dynamic regimes, varying in dependence on the intensity of interaction between the cells, are observed. The overall system is controlled by two parameters τ_F and D.

In our experimental arrangements, the one-directional coupling is relatively weak in comparison with the intensity of the mutual interaction between the nonautonomous and autonomous oscillators. Hence, the behaviour of the nonautonomous oscillator is strongly affected by the coupling. The types of regimes observed in the nonautonomous cell for different values of the intensity of interaction D between the cells are given in Table 2. The regimes are characterized by the values of experimentally evaluated rotation numbers ρ_{exp} defined as the ratio of the number of oscillations in the nonautonomous cell to the number of forcing pulses during an experiment run. The power spectra were used to differentiate between the types of the observed regime.

When the interaction between the cells was strong, then the synchronization of oscillatory regimes both in frequency and phase was

Table 2

$D \backslash \tau_F$	1.23	1.38	1.54	$D \backslash \tau_F$	1.85	2.07	2.30
0	P 1.00	P 1.00	A 1.41	0	P 1.50	P 2.00	P 2.00
6	P 1.00	A 1.22	A 1.67	10	P 2.00	A 2.30	P 2.50
8	A 1.28	A 1.44	P 2.00	15	A 2.12	A 2.54	P 3.00
10	P 1.50	A 1.52	P 2.00	20	A 2.05	A 2.86	P 3.00
15	P 1.50	P 2.00	P 2.00	A - aperiodic regime P - periodic regime			

observed. However, the synchronization can be destroyed even for strong couplings, when a pulse perturbing the nonautonomous oscillator occurs close before fast jump of the redox-potential. In this case, one oscillation of the perturbed oscillator is deleted while the autonomous cell oscillates regularly.

4. Conclusions

Studies of dynamic regimes in model nonlinear chemical systems may help to interpret complex transitions observed in various biological systems /9/ (synchronized cultural assemblies of excitatory cells in muscle and heart tissues, etc.).

References

1. C. Vidal, A. Pacault (Eds.): Nonlinear Phenomena in Chemical Dynamics (Springer, Berlin Heidelberg New York 1981)
2. H.L. Swinney, J.C. Roux: Non-equilibrium Dynamics in Chemical Systems (Ed.: C. Vidal, A. Pacault) (Springer, Berlin Heidelberg New York 1984)
3. I. Stuchl, M. Marek: Biophys.Chem. 3, 241 (1975)
4. I. Stuchl, M. Marek: J.Chem.Phys. 77, 1607, 2956 (1982)
5. M. Marek: in: Temporal Order (Ed.: L. Rensing, N.I. Jaeger) (Springer, Berlin Heidelberg New York 1984)
6. M. Dolnik, I. Schreiber, M. Marek: Phys.Lett. 100A, 316 (1984)
7. R.J. Field, M. Burger (Ed.): Oscillations and Travelling Waves in Chemical Systems (Wiley, New York 1985)
8. M. Dolnik, I. Schreiber, M. Marek: Periodic and Chaotic Regimes in Coupled and Forced Chemical Reactors, 8th CHISA Conf. Prague 1984
9. I. Cooke, M. Lipkin (Ed.): Cellular Neurophysiology (Holt, Rinehart and Winston, New York 1972)

Pattern in Marangoni-Instability Varied by Shear Stress Between Interface and Bulk

W. Waldhelm and H. Linde

Akademie der Wissenschaften der DDR, Zentralinstitut für physikalische Chemie, Rudower Chaussee 5, DDR-1199 Berlin, GDR

1. Introduction

Pattern formation and selection by selforganization is one of the most fascinating nonlinear mechanisms due to irreversible processes with internal feedback. Interfacial dynamic instabilities like Marangoni-instability show a surprising variety of different kinetic regimes and the corresponding pattern formating dissipative structures /1/, including patterns of stationary simple or hierarchically substructured systems, classical (mechanical) oscillations with wave interference patterns and relaxation oscillation with autowave patterns too. These stationary as well as moving patterns can easily be influenced and even extended by coupling with further different driving forces. For instance, the old spectacular results of Benard's experiments are caused by coupling of Rayleigh-instability with Ma-instability. That means, that the buoyancy forces in the unstable gradient of density are cooperating with interfacial tension forces by heat transfer and/or mass transfer of surface active matter. The last is easily realizable in two component liquids consisting of a quickly evaporating liquid of small surface tension and a slowly or not evaporating liquid respectively a soluted component with a higher surface tension. Heat transfer alone by heat conduction or in the consequence of evaporation cooling is also able to realize conditions of Ma-instability.

In this coupling Ma-instability often dominates in forming simple hexagonal patterns in thin layers /2/.

In the instability of the earth's atmosphere, we find coupling of Ra-instability, Taylor-instability and Coriolis forces at least.

The result is the extremely hierarchically substructured weather system with a lot of bifurcations and singularities too. As in other hydrodynamic instabilities with strong driving forces, a chaotic feature is superposed.

Recent experiments by KRINSKY and by KUHNERT /3,4/ show the possibility of coupling of the pattern of chemical autowaves with the pattern of hydrodynamic instabilities. There is also a wide field of combined eddy and wave patterns in liquid film flows /5/.

In this paper we inform about the variety of patterns, due to coupling of Ma-instability with the driving forces due to very slow laminar flow of thin liquid films.

2. Conditions and Methods of Measurement

To realize a slowly moving film which covers a solid plate, there are different possibilities, e.g.

1. by gravity at a perpendicular plate
2. by centrifugal force at a rotating horizontal plate
3. by shear stress of a forced convection on the plate
4. by MA-convection itself.

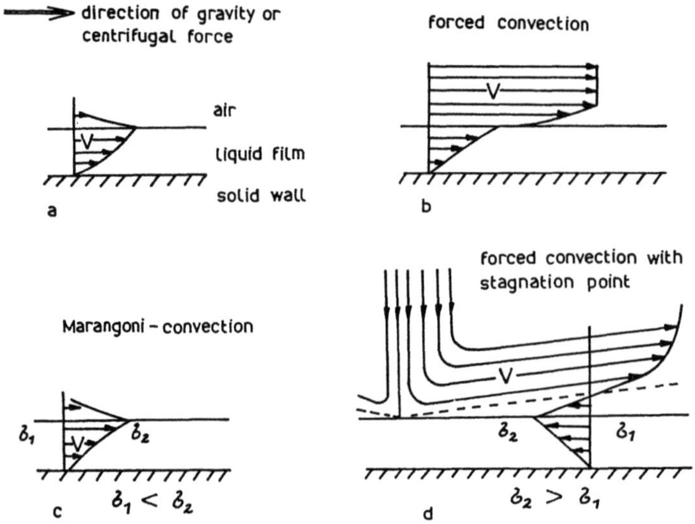

Fig. 1 Profiles of velocity v in film flow by gravity or centrifugal force (a), by forced convection (b), by Marangoni-convection (c) and by Marangoni-convection, induced by convective heat and/or mass transfer (d)

The first case occurs if sheetglass or sheetmetal is pulled out of a one- or two-component liquid and if the film is flowing down under the force of gravity (Fig. 1a). In case 2 we use a rotating plate covered with a film, inducing a centrifugal flow by the centrifugal force (Fig. 1a). The case 3 is given, if a forced convection is directed perpendicularly or tangentially to the thin liquid film (Fig. 1b). Case 4 with Ma-convection can be realized by a temperature gradient or concentration gradient of surface active matter along the surface (Fig. 1c). Even in condition 3 using e.g. a jet of air directed perpendicularly to an evaporating one- or two-component liquid, a countercurrent Ma-convection against the forced convection is realizable. It is a consequence of the surface tension gradient built by heat and mass-transfer (Fig. 1d).

In the big roll cells of second or higher order /2/ a long-range interfacial tension gradient is also working as a selfamplified internal Ma-convection with influence on the pattern of small roll cells of first order.

In order to visualize an image of the convection structure, we use very small films of lac, which dry quickly enough by evaporation of the used solvent. The convection structure is imaged in a structure of filmthickness as a consequence of the condition of continuity of flow and of friction at the wall. Therefore, a related thickness profile is fixed in the dry film. With aid of light diffraction by the curvature of film surface and/or by the phase contrast technique, we can easily show the two-dimensional pattern of this convection structure.

3. Discussion of Experimental Results

Though the velocity gradients in these different film flow conditions are rather different, we observe in principle the same qualitative influence of film flow in the pattern of Ma-instability. That means, we can produce in principle the analog pattern with film flow by gravity, by centrifugal force, by forced convection, by long-range Ma-

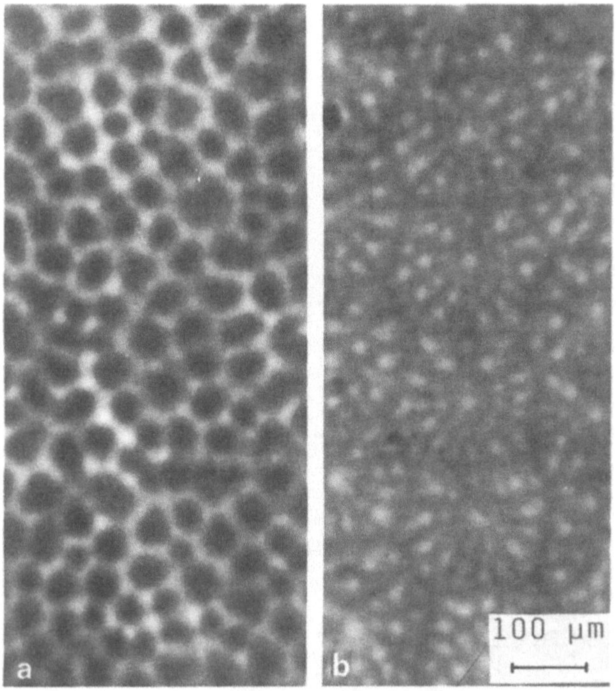

Fig. 2 Hexagonal pattern of roll cells, obtained without flow of the entire film (a - roll cells of first order, b - roll cells with substructure)

convection and even if internal Ma-convection takes place in roll cells of higher order. Starting without film flow (Re = 0), we reproduce at very thin and very quick solidifying films of an initial thickness of about 10 µm the well-known simple hexagonal patterns of roll cells of first order only, which are typical for Ma-instability (Fig. 2a).

That means, that at very quickly solidifying thin films the pattern corresponding to the initial thickness is only fixed at rest. If the initial film thickness is $\gtrsim$ 30 µm, we observe firstly hexagonal roll cells of greater size corresponding to the initial thickness. They are later, with decreasing film thickness because of evaporation of the solvent (the lac contains about 99.5% solvent!) fulfilled with smaller roll cells corresponding to the final thickness of the still fluid film (Fig. 2b). With increasing film flow velocity ($0 < Re < 1$), the tendency of formation of polygonal pattern is overlapped with the tendency of a longitudinal orientation of this basic pattern (Fig. 3). In these results we have examples of the consecutive working of two rather varied selforganization processes of dissipative structures, which is expected to be a fundamental principle in morphogenesis (Figs. 2b, 3, 4, 5):
Firstly, at a greater thickness and a greater Re-number, the longitudinal orientation with greater wavelength is organized, later, at a smaller thickness and Re-number, the tendency of amplification of small-sized polygonal pattern is working additionally and simultaneously. In pictures like Figs. 3 and 4 we find longitudinal stripes with an about 2 to 13 times greater wavelength in comparison with the small roll cells. The stripes form in thin liquid films firstly a rather square structure. In thick films, we find a rather square superstructure in which the small cells realize consecutively dif-

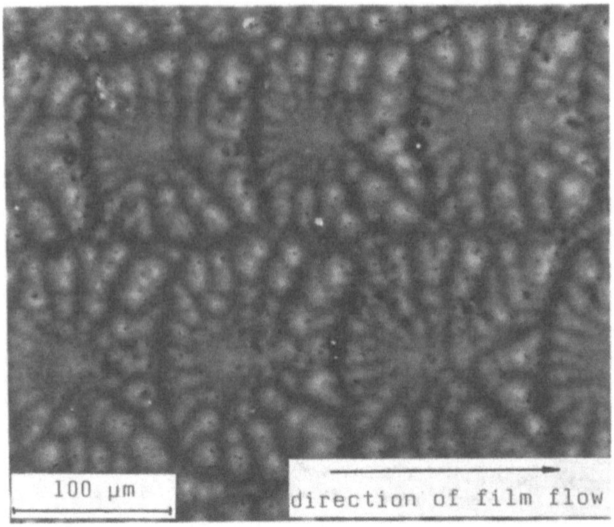

Fig. 3 Roll cells with substructure: longitudinal orientation of the basic pattern due to the film flow

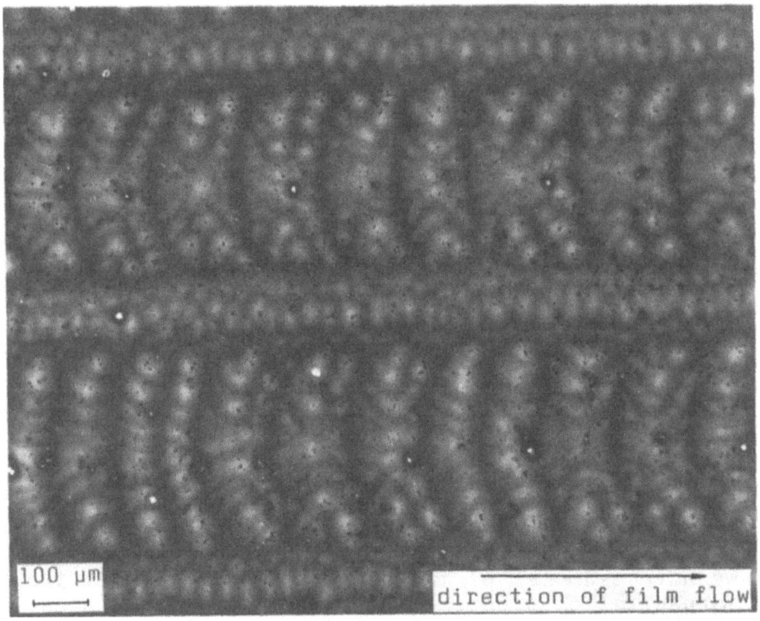

Fig. 4 Ladder structure substructured in beams and rungs

ferent kinds of centrosymmetric substructures. With increasing flow velocity of films of about 50 µm initial thickness a ladder structure with the orientation of the roll cells in beams and rungs of ladders appears. The rungs are formed by aligned roll cells. Even the beams can occasionally be formed by one to three lines of aligned roll cells. There are a lot of different transition structures between the reported idealized patterns. As already remarkable in Fig.

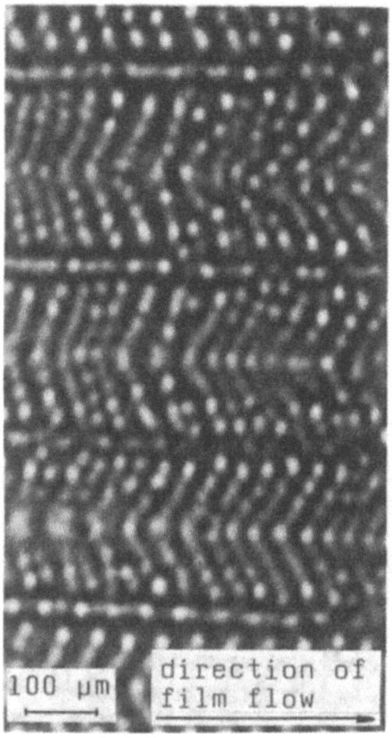

Fig. 5 Substructured chevron structure

4, we find also a tendency to force chevron structures with further increasing flow velocity (Fig. 5).

A chevron structure is also remarkable in the aligned cells of the rungs, from which we get the information of a flow direction component along the rungs to the beams. Also pure beam structures (stripes) are observed. They correspond to low Ma-number and low Re-number and transits in a ladder structure, if the flow velocity of the forced convection increases. Chevronlike and centrosymmetric substructures show a still internally flowing superstructure during origination of small roll cells.

4. General Discussion

As in coupling of Ma- and Ra-instability, mutual lowering of critical conditions is expected according to the scheme Fig. 6 with the Ma-number of heat or mass transfer (stationary roll cell regime),

$$Ma_1 \equiv - \frac{d\sigma}{dT} \frac{d^2 \cdot \beta_T^{(1)}}{\lambda^{(1)} \cdot \eta^{(1)}}$$

(σ - surface tension; T - temperature; β_T - gradient of the temperature; d - unit length; λ - thermal diffusivity; η - dynamic viscosity)

and

$$Ma_1 \equiv - \frac{d\sigma}{dc} \frac{d^2 \cdot \beta_c^{(1)}}{D^{(1)} \cdot \eta^{(1)}}$$

[σ - surface tension; c - concentration; β_c - gradient of the concentration; d - unit length; D - coefficient of diffusion; η - dynamic viscosity (index 1 marks fluid 1)]

and the Re-number

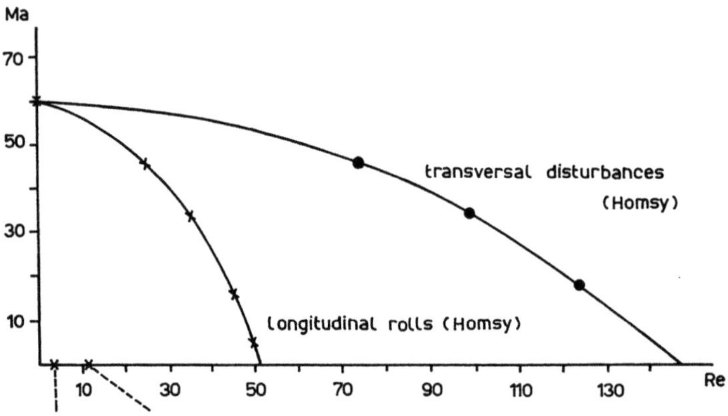

Fig. 6 Interaction between Ma-instability (stationary regime) and shear-stress-driven longitudinal rolls and transversal disturbances after HOMSY and GUMERMAN. See also $Re_{critical}$ for laminar transversal waves at Re = 4 and 12 after /5/

$$Re = \frac{v \cdot d}{\nu}$$

v - velocity of the surface for the mainflow; d - thickness of the fluid film; ν - kinematic viscosity of the flowing film after HOMSY and GUMERMAN /6/.

The nonlinear theory /6/ is able to distinguish between the amplification of longitudinal rolls and transversal disturbances. Because of the longitudinal orientation in the above-discussed patterns and the lower critical Re-number of about 50 for the longitudinal rolls, we prefer to assume a coupling of Ma-instability with this longitudinal mode, though the Re-numbers of our experiments are much smaller with $Re \leq 1$. Maybe the consideration of more realistic assumptions in the theory gives free play to lowering the critical Re-number for longitudinal rolls. We guess to work in the unstable region not far from stable region. The reason for this conclusion is, that we observe only patterns of rather high order without a dominant time-dependent behaviour. (Time-dependent behaviour itself, i.e. propagating roll cells in a hierarchical superstructure and relaxation oscillation of roll cells of higher order /1/ and irregularities in spatial and time-dependent behaviour are indications for strong exceeding of critical conditions for instability.)

Of course the rather high order is often disturbed by dislocations (Fig. 7).

From experiments and theory /5/, it is further well known that transversal laminar waves occur at $Re > 4$ or 12 and transit into turbulent transversal waves in the range of $Re \sim 500$. The above-mentioned Re-numbers of our experiments are lower than these values and therefore wavy instabilities are also hereby not observed.

The here-reported results can be discussed as an indication of the tendency to formation of longitudinal rolls of Ma-instability by cooperation with shear stress as the driving force for instabilities. It is of course a further aim to coordinate quantitatively the kinetic and topological behaviour of the resulting dissipative structures to the field of system properties at least under distinguishing between Ma-numbers for different regimes /6/ and in addition to further system properties like Pr-number e.g. too.

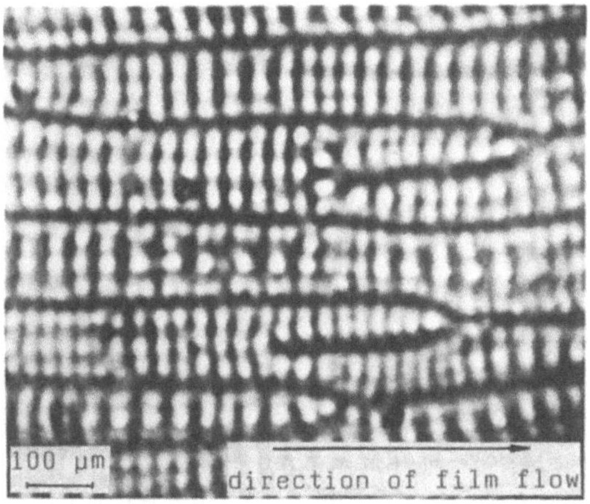

Fig. 7 Dislocations in a ladder structure

Acknowledgement

Thanks to Dr. Lothar Weh of our institute for imaging Figs. 5, 6a, 6b in phase contrast technique.

References

1. H. Linde: Time-dependent Behaviour and Regularity of Dissipative Structures of Interfacial Dynamic Instabilities, Faraday Discuss.Chem.Soc. 77, 181 (1984)
2. E.L. Koschmieder: Old and New Problems in Benard Convection; Dissipative Structures of Marangoni-Instability, in: Convective Transport and Instability Phenomena (Ed.: J. Zierep, H. Oertel jr., G. Braun, Karlsruhe 1982)
3. K.I. Agladze, V.I. Krinsky, A.M. Pertsov: Chaos in the non-stirred Belousov-Zhabotinsky reaction is induced by interaction of waves and stationary dissipative structures, Nature 308, 834 (1984)
4. L. Kuhnert, H.J. Krug, L. Pohlmann: Physica D submitted
5. Ch. Boyadjiev, V. Beschkov: Mass Transfer in Liquid Film Flows (Publishing House of the Bulgarian Academy of Sciences, Sofia 1984)
6. J. Reichenbach, H. Londe: J.Colloid Interface Sci. 84, 433 (1981)
7. R.J. Gumerman, L.M. Homsy: AICHE Journal 20, 981, 1161 (1974)

Ostwald Ripening in Viscoelastic Media

J. Schmelzer[1] *and I. Gutzow*[2]

[1]Wilhelm-Pieck-Universität Rostock, Sektion Physik, Universitätsplatz 3, DDR-2500 Rostock, GDR
[2]Institute of Physical Chemistry, Bulgarian Academy of Sciences, BG-1040 Sofia, Bulgaria

Abstract

The theory of Ostwald ripening in elastic media proposed earlier is extended to include the influence of viscous properties on the course of Ostwald ripening. Viscous properties are taken into account by a special model discussed in detail elsewhere.

Phase transformations in solids are often accompanied by the development of elastic strains due to the formation and growth of clusters of a new phase /1/. In particular, the development of elastic strains modifies the kinetics of crystallization, the kinetics of growth of single clusters and the process of Ostwald ripening in glassforming melts /2,3/, which can be considered as a special case of viscoelastic media. The aim of the present paper consists in the investigation of the influence of viscous properties on the process of Ostwald ripening in viscoelastic media.

Viscoelastic media are characterized both by elastic and viscous properties. If the elastic strains developing during the growth of a cluster are due to the difference in the volume per particle in the different phases, the total energy of elastic deformations $G^{(\varepsilon)}$ can be expressed by

$$G^{(\varepsilon)} = \varepsilon V_\alpha , \quad \varepsilon = \text{const.} \tag{1}$$

V_α being the volume of the cluster. Though this expression was first obtained in an investigation of elastic strains developing during allotropic transitions /7/ it is widely used also for the description of elastic strains in solid solutions considered here /1/. It has, however, to be taken into account that the application of (1) to solid solutions presumes that the particles of the different phases change their places during the growth of the clusters. Therefore, implicitly a transport of mass over finite distances and consequently viscous properties are presumed. Segregation processes in purely elastic media lead to elastic strains which can be described by (2) /6/:

$$G^{(\varepsilon)} = \varkappa \frac{(V_\alpha - V_0)^2}{V_0} \theta \qquad \theta = \begin{cases} 0, & V_\alpha \leq V_0 \\ 1, & V_\alpha > V_0 \end{cases} \tag{2}$$

$V_0(R_0)$ has the meaning of an initial volume (radius) of the clusters, for which strains of this type begin to develop. The parameters ε and $\varkappa$ can be expressed by the elastic constants of the medium. While elastic strains of the type (1) only modify the growth rate quantitatively, strains of the form (2) lead to a stop of the growth for $V_\alpha > V_0$.

Viscous properties of the medium can be explicitly taken into account by a model discussed in detail elsewhere /6/. In a first approximation this model leads to an equation of the form (2) again

$$G^{(\varepsilon)} = \varepsilon V_\alpha + \varkappa' \frac{(V_\alpha - V_0)^2}{V_0} \theta , \tag{3}$$

where $\varkappa'$ can be expressed by

$$\varkappa' = \frac{\varkappa \left(\frac{C_\alpha}{C} - 1 \right)}{\left[1 + \frac{R_0 (\sqrt[3]{C_\alpha / C} - 1) + \sqrt{D_M t}}{R_0} \right]^3 - 1} \; ; \tag{4}$$

c_α and c are the concentrations of the segregating particles in the cluster and in the homogeneous initial state. Viscous properties of the medium are taken into account by the selfdiffusion coefficient D_M of the matrix particles. If D_M tends to zero, $\varkappa'$ is equal to $\varkappa$ again. For a high value of the diffusion coefficient, $\varkappa'$ is small and elastic strains of the type (2) do not influence the growth. Equation (3) leads to the following differential equation for the radius R_α of a single cluster:

$$\frac{dR_\alpha}{dt} = \frac{DC_\beta}{c_\alpha kTR_\alpha} \left\{ \mu_\beta - (\mu_\alpha + \frac{\varepsilon}{c_\alpha}) - \frac{2\sigma}{c_\alpha R_\alpha} - \frac{2\varkappa'(V_\alpha - V_0)}{c_\alpha V_0} \theta \right\} . \tag{5}$$

D is the diffusion coefficient, c_β the variable concentration of the segregating particles during the growth, μ their chemical potential in the cluster (α) and the medium (β), k the Boltzmann constant and σ the surface tension.

Considering the solid solution as an ideal mixture, (5) was solved numerically for different values of the diffusion coefficient D_M. In dependence on the actual value of D_M ($D_M^1 = 4.3 \cdot 10^{-15} ms^{-2}$; $D_M^2 = 0.1\ D_M^1$; $D_M^3 = 0.01\ D_M^1$; $D_M^4 = 0.001\ D_M^1$), growth between the limiting cases $\varkappa' = 0$ and $\varkappa' = \varkappa$ are obtained (Fig. 1). c' is the equilibrium concentration at a planar interface. Generally, in a nucleation period a big number of clusters is formed. Their growth leads to a depletion

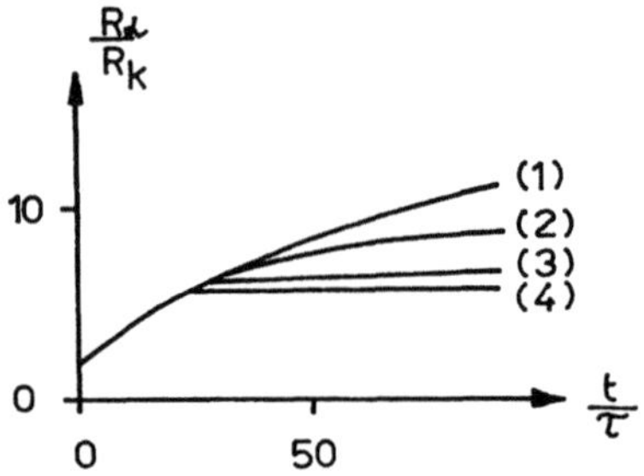

Fig. 1 Growth of a single cluster in a viscoelastic medium ($\varepsilon = 2.37\cdot 10^7 Nm^{-2}$; $\varkappa = 4.8\cdot 10^8 Nm^{-2}$; $\sigma = 0.08 Nm^{-1}$; $D\delta = 1.2\cdot 10^{-5} m^2 s^{-1}$; $T = 730$ K; $\tau = 1.62\cdot 10^{-2} s$; $\delta = 8/27$; $R_k = 6.28\cdot 10^{-10} m$; $c' = 1.4\cdot 10^{26} m^{-3}$; $c = 4.2\cdot 10^{26} m^{-3}$; $c_\alpha = 2.3\cdot 10^{28} m^{-3}$; $R_o = 5.2 R_k$)

of the surrounding medium and a competitive growth denoted as Ostwald ripening. In preceding papers /4,5/ a new method of kinetic description of this process has been proposed. The equations describing the development of the mean radius of the clusters R and the number of clusters s during the process of Ostwald ripening according to our theory are given by:

$$\frac{dR}{dt} = \frac{c_\beta D\delta}{c_\alpha^2 kTR^2}\left\{\sigma + \frac{3}{4\pi R^2}G^{(\varepsilon)} - V_\alpha \frac{\partial G^{(\varepsilon)}}{\partial V_\alpha}\right\}\frac{1}{Z}\left\{1 + Z - \frac{R^2}{2\sigma}\frac{\partial^2 G^{(\varepsilon)}}{\partial R\,\partial V_\alpha}\right\},$$

$$\frac{d}{dt}\left\{\ln\left(\frac{sV_\alpha}{V}\right)\right\} = -\frac{1}{Z}\left\{1 - \frac{R^2}{2\sigma}\frac{\partial^2 G}{\partial R\,\partial V_\alpha}\right\}\frac{d}{dt}\left\{\ln(R^3)\right\}, \tag{6}$$

$$Z = 3\frac{c_\alpha kTR}{2\sigma}\frac{V_\alpha s}{V}\left\{-\frac{c_\alpha}{c(1 - s\frac{n_\alpha}{n})} + \frac{1}{(1 - s\frac{V_\alpha}{V})} + \frac{1}{(1 - s\frac{V_\alpha}{V})^2}\right\}.$$

V_α is here the mean volume of a cluster, V the volume of the whole system and n the total number of the segregating particles in the system.

In analogy to the growth of single clusters, elastic strains lead to a qualitative change of the growth process only if elastic strains grow more rapidly than linearly with the volume of the cluster.

In the Figs. 2-7, results of numerical calculations are presented for the mean radius R, the mean volume V_α, the number of clusters s, the total volume V_{ov} and the total surface area O_{ov} of the new phase. The process of growth can be divided into three stages. In the first stage, the total volume of the new phase grows rapidly and the quantity Z^{-1} tends to zero. The second stage is governed by the equations $V_\alpha \sim t$ and $s \sim t^{-1}$, derived first by LIFSHITZ and SLYOZOV /8/. Elastic strains of the type (2) lead to a decrease of the growth rate. The system of clusters tends to a state corresponding to an ab-

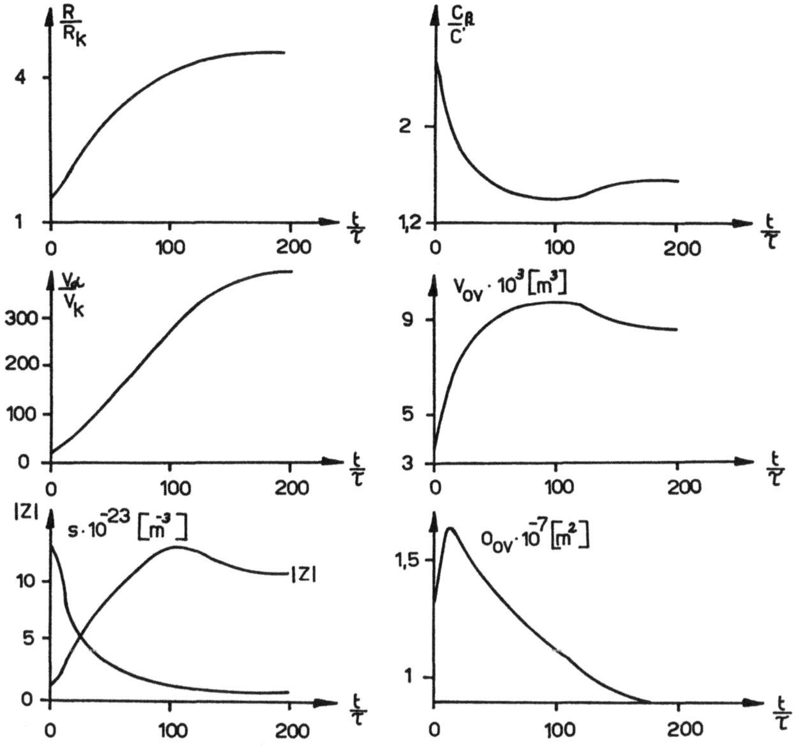

Figs. 2-7 Ostwald ripening in viscoelastic media ($R_0=4.2R_k$; $D_M=4.3\cdot10^{-17}m^2s^{-1}$, for the value of the other parameters see the legend to Fig. 1)

solute minimum of the thermodynamic potential. The further slow growth of the clusters is due to a change of this position connected with the relaxation of the elastic strains. By a variation of D_M, growth rates for the mean cluster radius between the limiting cases $\varkappa'=0$ and $\varkappa'=\varkappa$ can be obtained again (Fig. 8).

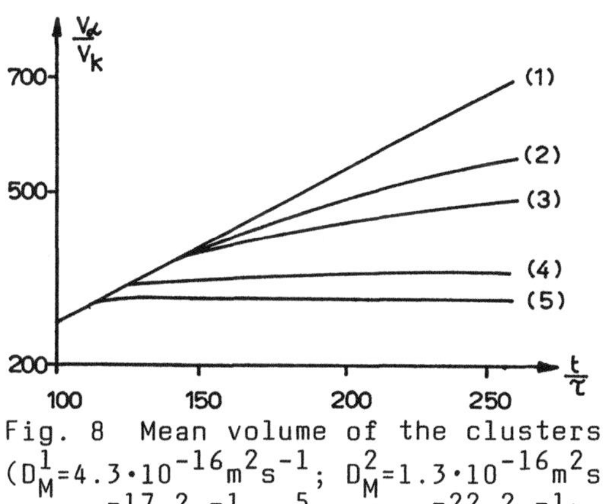

Fig. 8 Mean volume of the clusters for different values of D_M ($D_M^1=4.3\cdot10^{-16}m^2s^{-1}$; $D_M^2=1.3\cdot10^{-16}m^2s^{-1}$; $D_M^3=7.3\ 10^{-17}m^2s^{-1}$; $D_M^4=1.3\cdot10^{-17}m^2s^{-1}$; $D_M^5=4.3\cdot10^{-22}m^2s^{-1}$)

References

1. J.W. Christian: The theory of transformation in metals and alloys (Oxford 1975)
2. I. Gutzow, J. Schmelzer, R. Pascova, B. Popow: Rostocker Phys. Manuskr. 8, 22 (1985)
3. R. Pascova, I. Gutzow: Glastechnische Berichte 56, 324 (1983)
4. J. Schmelzer: Z.Physik.Chem. (Leipzig) 266, 1057 (1985); 266, 1121 (1985)
5. J. Schmelzer, I. Gutzow: On the kinetic description of Ostwald ripening in elastic media, submitted for publication
6. J. Schmelzer, I. Gutzow: On the theoretical description of the growth of a new phase in elastic and viscoelastic media, in preparation
7. F.R.N. Nabarro: Proc.Phys.Soc. 52, 90 (1940); Proc.Roy.Soc. A175, 519 (1940)
8. I.M. Lifshitz, V.V. Slyozov: J.Phys.Chem.Sol. 19, 35 (1961)

Thermodynamic Parameters and Structural Criteria for Phase Transformations in Amorphous Covalent Bound Materials

W. Pompe and A. Richter

Zentralinstitut für Festkörperphysik und Werkstofforschung der Akademie der Wissenschaften der DDR, Helmholtzstraße 20, DDR-8027 Dresden, GDR

1. Introduction

Crystallization studies are of interest not only for examining the fundamentals of the involved nucleation and growth processes in some detail, but also for learning to produce novel microstructures by controlled crystallization. Experimental and theoretical investigations of crystallization of amorphous solids show that the formation of metastable crystals occurs as transient or final stages /1,2,3/. These general considerations can be applied to experimental investigations of amorphous carbon films. With a laser plasma, carbon films are deposited on different substrate materials /4/. Depending on deposition conditions, but also as the result of a defined thermal treatment of these layers, diamond and various carbines are obtained as crystalline particles within the amorphous matrix. A similar behaviour can be observed for other carbon films deposited by high-energy beam techniques /5/.

The general question analyzed in this paper is, how structural peculiarities determine the kinetics of formation of critical nuclei and their growth rates during partial crystallization of different phases. Spacial emphasis is put on the role of stored elastic energy and the influence of interfacial energy on steady-state and transient nucleation within the context of classical nucleation theory /6/ applied to small crystalline particles embedded in a solid matrix.

Considering different mechanisms of structural transformations, two possible ways are discussed:

(i) The first one may be called "thermodynamically controlled nucleation" that contains crystallization under elastic constraints to such an extent that the metastable phase is the phase with the minimum of free energy

$$F_{met}(Z_n(t \to \infty); \Phi) < F_{stab}(Z_n(t \to \infty); \Phi) \, . \qquad (1)$$

$Z_n(t \to \infty)$ is the probability of the existence of a critical nucleus with size n at time $t \to \infty$ and Φ is a set of constraints explained later.

(ii) The second one will be termed "kinetically controlled nucleation" that means crystallization under conditions where the probability $Z_n^{met}(t)$ for the existence of a metastable nucleus of size n is larger than the corresponding probability of the stable phase $Z_n^{stab}(t)$ though condition (1) is not fulfilled

$$Z_n^{met}(t) > Z_n^{stab}(t) \, . \qquad (2)$$

However, subsequent transformation into the thermodynamically stable phase has to be prevented by freezing-in the obtained structure.

2. Theoretical Model

Applying the well-known Becker-Döring theory to solid-solid transformations, calculations of changes in the free energy between different phases have to be performed:

$$\Delta F = \Delta F_{ph} + \Delta F_{el} + F_{\gamma} \quad . \tag{3}$$

ΔF_{ph} describes the volume free energy change to form from a phase α the quantity β without internal mechanical stresses. Additional contributions ΔF_{el} to the volume part of the free energy arise from elastic deformations within the amorphous matrix itself, but also due to elastic interactions between the crystalline nucleus and the amorpous matrix. F_{γ} represents the internal boundary energy corresponding to the surface term in nucleation theory. Liquid phases as intermediate states during structural changes are excluded from the present considerations.

2.1. Amorphous Matrix

The atomic structure of non-crystalline carbon contains not only topological disorder of atoms, but variation in bond character, too, that is a mixture of both sp^2 and sp^3 hybridized atoms /7/. The bond structure also characterizes crystalline phases as carbines (sp), graphite (sp^2) and diamond (sp^3). Therefore, the introduction of an effective structural parameter would offer a good approximation to estimate the free energy contribution ΔF_{ph}. Taking into account phonons as a sensitive measure for local atomic bond characteristics, and assuming this contribution to be the dominant one, the Einstein temperature Θ_{α} determines the free energy of the phase α as a structural parameter in a first approach

$$F(\Theta_{\alpha}) = 3\,NkT\,\left[\frac{1}{2}\,\frac{\Theta_{\alpha}}{T} + \ln\,(1-\exp\,(-\,\Theta_{\alpha}/T))\right] \quad . \tag{4}$$

The Einstein temperature for structures with covalent bonding can be estimated with the binding energy U_b and the nearest neighbour distance r_o with

$$\theta = \frac{k}{\hbar}\left[\frac{27}{r_o^{\,2}}\,U_b\,\,z\,\langle\alpha\rangle\right]^{1/2} \quad , \tag{5}$$

where k is Boltzmann's constant and $\hbar$ Planck's constant, z is the average coordination number and the factor $\langle\alpha\rangle$ describes the spatial distribution of lattice vibrations. Amorphous carbon contains the fraction x sp^3 bonds and the fraction (1-x) sp^2 bonds. Different amorphous carbon structures correspond to Einstein temperatures in the range of Θ_{α_1} = 1862 K and Θ_{α_2}= 1558 K. Using formula (5) for the limiting case of crystalline phases, too, calculated Einstein temperatures agree well with experimentally determined Debye temperatures /8/. The free energy difference of the amorphous phase compared to the crystalline carbon structures, diamond and graphite, is given by

$$F_{ph} = F(\theta_a) - F(\theta_c) \quad , \tag{6}$$

where Θ_a and Θ_c are the Einstein temperatures of the amorphous and crystalline phase, respectively.

An additional contribution to the free energy arises from topological disorder of the amorphous solid described by a defect network /9,10/. Structural elements of amorphous covalent bound materials are identified with defects, mainly disclinations /11.12/. Thus, the ring structure of amorphous carbon can be modelled by a distribution of wedge disclinations with positive and negative signs of rotation angle ω. Analysis of experimental investigations of amorphous carbon enables the estimation of the disclination density in the range ϱ = 0.3 ... 0.6. Moreover, there exists an asymmetry between five- and sevenfold rings $a = |\varrho^- - \varrho^+|/\varrho = 0.25$, where $\varrho^\pm$ are the partial disclination densities for positive and negative rotation angles. Such a defect network contains stored energy from elastic strains. Each structural unit has energy contributions from the self-energy E_i^s and interactions W_{ij} with all other defects j in the surrounding defined by a cut-off radius r_{cl}

$$E_i = E_i^s + \frac{1}{2} \sum_{j \in U_i} W_{ij} \quad . \tag{7}$$

Such a cluster defect structure with internal interfaces built up with screened disclinations in a non-balanced distribution ($a \neq 0$) corresponds to a dipole distribution of simple non-screened disclinations ($a=0$). The average elastic strain energy E of the defect network has been estimated as 3.3 and 4.5 kJ mol^{-1} for soft and hard amorphous carbon layers, respectively, using different elastic moduli for the material considered.

Applying cluster analysis to the defect network /10/, structural inhomogeneities are detected characterized by fluctuations of the average stress and energy of the structure. Strongly strained clusters contain a high amount of elastic strain energy and represent prestages of crystalline nuclei. The size of these patterns ranges between 15 and 35 Å in diameter, the relative volume fraction compared to the whole material is about 15%. These highly strained defect clusters with local energies of 22 ... 30 kJ mol^{-1} can be regarded as "weak spots", which can be transformed with low activation energies in crystalline particles.

The free energy of the amorphous phase compared to the crystalline carbon structures, diamond and graphite, is thus given by

$$\Delta F = F(\theta_a) - F(\theta_c) + \Delta F_{el}(\varrho) \ , \tag{8}$$

where ΔF_{el} results from the elastic strain energy of the frozen-in defect structure. Without external pressure the ground state of carbon is graphite for all temperatures. Since all amorphous states lie energetically above graphite, phase transformations into graphite are always possible. Diamond represents the metastable carbon phase under normal conditions. Preferably hard amorphous carbons with high Einstein temperatures can be transformed into diamond by thermodynamically controlled nucleation processes.

2.2. Partially Crystallized Solid

Small solid particles in a solid amorphous matrix cause contributions to the free energy. Due to interfaces between the two phases, a boundary free energy F_γ has to be taken into account. The interface energy has been estimated within an atomistic model in terms of nearest-neighbour bond energies /13/. In case of covalent bound materials, the surface free energy contains an energy contribution of unsaturated bonds $F_{\gamma,f}$ and another one from a misfit between bonds of different phases $F_{\gamma,m}$. Local coordination numbers enter dominantly in the calculation of the number of bonds per internal boundary and characterize the crystalline as well as the amorphous phases. In Fig. 1, surface energies for graphite and diamond nuclei in an amorphous carbon matrix in dependence on its average coordination number z_a are shown. The parameter $\varkappa$ is the ratio between the fraction of sp^3 bonds and that of the sum of sp^2 and dangling bonds /14/.

Growth of crystalline nuclei is connected with an elastic misfit between the crystalline phase and the amorphous matrix, which causes

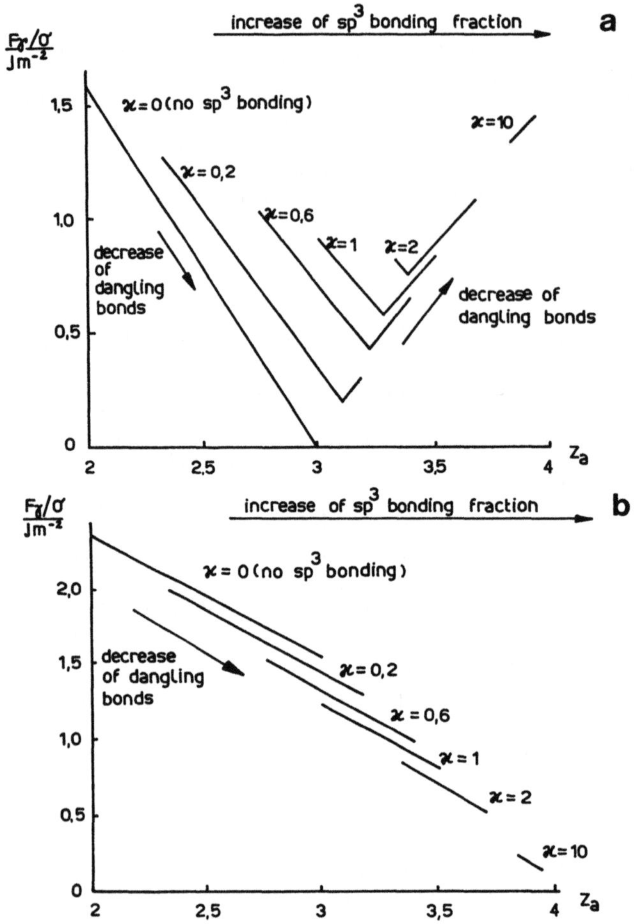

Fig. 1 Boundary energy of graphite (a) and diamond (b) nuclei in an amorphous matrix with the average coordination number z_a (σ - surface of the nucleus)

an elastic contribution ΔF_{el} to the free energy /14/. For a given stress-free deformation

$$\varepsilon^T_{ij} = e^T \delta_{ij} + e^T_{ij} \tag{9}$$

denoting isotropic dilatation or compression e^T and shear deformation e^T_{ij}, the density of the elastic energy u can be calculated in dependence on the volume fraction of nuclei and effective shear and compression moduli G* and K*, respectively. The effective quantities can be estimated within Hashin's approximation /15/ in the case of a distribution of elastic isotropic nuclei. The elastic energy density u contains contributions from isotropic deformation, shear deformation and, according to forces between the thin deposited layer and the substrate, contributions of a fixed macroscopic external stress

$$u = f(V_n, K_i, G_i)(e^T)^2 + f(G_i, G^*, \gamma^*)(\tfrac{1}{2}\, e^T_{ij}\, e^T_{ij})$$
$$+ f(K^*, G^*)(\varepsilon_0 - \varepsilon^*)^2 \; . \tag{10}$$

V_n is the volume fraction of nuclei, K_i and G_i are compression and shear moduli of the phase i, respectively, γ^{*1} is the effective Poisson ratio and ε_o and ε^* are the deformation according to an external stress σ_o and the deformation in the stress-free state of the heterogeneous layer system, respectively. The deformation e^T can be estimated by changes of the mass density between the two phases

$$e^T = \frac{1}{3}\,\frac{\rho_a - \rho_c}{\rho_c} \quad , \qquad (11)$$

where ρ_a and ρ_c are the mass densities of the amorphous and crystalline phase, respectively. The elastic free energy density depends strongly on this material parameter. Moreover, the used elastic constants for the amorphous matrix are dependent on the sp^3 bonding fraction related to the mass density. In order to regard the influence of very high stresses, it has to be assumed that the moduli of graphite increase with pressure /17/. The third term in (11) cannot be neglected, since it gives energy contributions of the strained heterogeneous system in the same range as those of the transformation amorph-crystalline without external stresses. In Fig. 2 /16/ the change of the elastic energy with the volume fraction of crystalline nuclei is plotted versus the density of the amorphous carbon layer in dependence on the external pressure σ_o. For high mass densities of the matrix, the necessary elastic energy change for crystal growth of diamond nuclei is smaller than that for graphite, and is even lowered for increasing external stress. Applying these results to the calculation of the critical activation energy for a nucleus, two characteristic regions of formation of phases are obtained, depending on the density of the layer and the initial stress. Diamond will preferably grow in amorphous films with a high mass density and at high pressure by thermodynamically, controlled growth /18/.

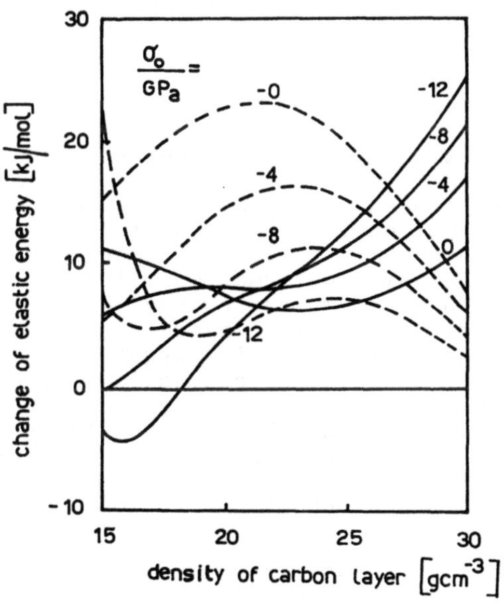

Fig. 2 Change of the elastic energy during nucleation in an amorphous carbon layer depending on its mass density and external stresses; graphite (———) and diamond (— — —) nucleation

3. Kinetically Controlled Phase Transitions

According to different stages of nucleation and growth of nuclei, characteristic times can be distinguished /19/:

- The average steady-state nucleation time

$$t_s = I_s^{-1} = \left\{ N\ R\ n_c^{\beta} \left[\frac{F_c}{kT} \frac{1}{3\pi n_c^{2}} \right]^{1/2} \exp(-F_c/kT) \right\}^{-1} , \qquad (12)$$

- the time lag of establishing the steady-state distribution of nuclei

$$\tau = \frac{0.32\pi}{R} \frac{n_c^{2-\beta}}{\beta F_c/kT} , \qquad (13)$$

- the growth time

$$t_g = n_{max}^{(1-\beta)} \frac{n_c}{\beta R F_c/kT} . \qquad (14)$$

I_s is the stationary nucleation rate, k is Boltzmann's constant, N is the number of nucleation centres. The size of the critical nucleus is given by

$$n_c = \left(\frac{\beta \gamma}{\delta f} \right)^{\frac{1}{1-\beta}} \qquad (15)$$

and the critical free energy by

$$F_c = \frac{1-\beta}{\beta} \delta f\ n_c , \qquad (16)$$

if $\Delta F = -\delta f\ n + \gamma n^{\beta}$ contains volume and surface contributions corresponding to formula (3). β can take two values, $\beta = 2/3$ for isotropic nucleus growth and $\beta = 1/2$ for anisotropic two-dimensional growth. R denotes a characteristic frequency of local mass transport R = $\gamma_o \exp(-U/kT)$ where U is the activation energy for the transport of an atom across the interface of a nucleus. The maximum nucleus size can be approximated by /20/

$$n_{max}^{-1} \simeq I_s \cdot t_g \frac{1-\beta}{2-\beta} . \qquad (17)$$

The different relations of characteristic times in dependence on supersaturation, critical nucleus size and free energy of the critical nucleus allow estimations on time development of phase formation. For thermodynamically controlled transitions, nucleation is a steady-state process and growth is determined by high values of supersaturation δf obtained by constraints, e.g. elastic deformations. Therefore, the final phase is the phase with minimum free energy. Kinetically controlled transitions characterize nucleation processes that deviate from stationary nucleation. The growth time is larger than the steady-state nucleation time t_s, since critical activation energies for nucleation are small. This means, if the boundary energy of the metastable crystalline phase is smaller than that of the stable crystalline phase, metastable crystalline nuclei can be formed by kinetically controlled nucleation processes (Fig. 3).
In order to distinguish these two main processes of nucleation, the following characteristic can be given:

thermodynamically controlled nucleation	kinetically controlled nucleation
$t_s/\tau > 1$	$t_s/\tau < 1$
$t_s/t_g > 1$	$t_s/t_g < 1$

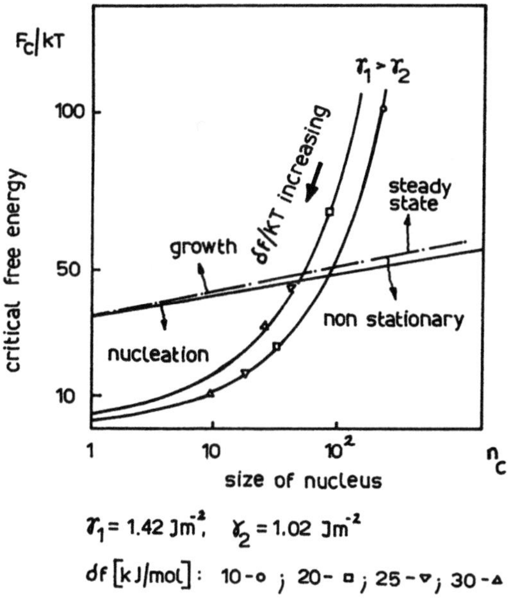

Fig. 3 Schematic representation of nucleation and growth processes; the two straight lines distinguish between areas of thermodynamically and kinetically controlled nucleation, full line separates $t_s/t_g \gtrless 1$, dotted line $\tau/t_s \gtrless 1$; the two other lines relate the critical free energy to the corresponding nucleus size; with increasing δf a change from thermodynamically to kinetically controlled processes can be stated for constant values of the boundary energy.

The evolution of the nucleus size distribution with time has been calculated in more detail from the Master equation.

In conclusion, we state that the classical nucleation theory has been successfully applied to crystallization processes of amorphous carbon. Solid-solid phase transformations have been considered within both thermodynamically and kinetically controlled nucleation. The elastic constraints change the free energy locally to such an extent that the metastable phase becomes the phase of the absolute minimum of the free energy within the context of thermodynamically controlled transitions. Kinetically controlled nucleation processes deviate from stationary nucleation and are characterized by low critical activation energies and low boundary energies. Thus, kinetically controlled transitions can produce metastable phases by nucleation during the transient stage, provided that subsequent transformation into the stable phase is prevented by an appropriate quenching regime.

Acknowledgement

We are indebted to our colleagues A. Bahr, W. Kreher and H.-J. Weiß for many helpful discussions.

References

1. U. Köster, U. Herold: in: "Glassy Metals I" (Eds.: H.-J. Güntherodt and H. Beck) (Springer, Berlin Heidelberg New York 1981)
2. I. Gutzow, I. Avramov: Thin Solid Films 85, 203 (1981)

3. H. Vogel, M.v. Heimendahl: Mater.Sci.Eng. 57, 171 (1983)
4. W. Pompe, H.-J. Scheibe, H.-D. Bauer, A. Richter, H.-J. Weiß, S. Völlmar, R. Krumphold, W. Brunner, P. Nicklis: Wiss.Ber. ZFW Nr. 26, 87 (1984)
5. C. Weißmantel, K. Bewilogua, K. Breuer, D. Dietrich, U. Ebersbach, H.-J. Erler, B. Rau, G. Reiße: Thin Solid Films 96, 31 (1982)
6. R. Becker, N. Döring: Ann.Physik (Leipzig) 24, 719 (1935)
7. A. Richter, W. Pompe: ZfK Ber. 555, 629 (1984)
8. V.B. Federov, M.Ch. Schorschorov, D.K. Chakimova: Kohlenstoff und seine Wechselwirkungen mit Metallen (Mir, Moscow 1978) (in Russian)
9. A. Richter, A.E. Romanov, W. Pompe, V.I. Vladimirov: phys.stat. sol.(b) 122, 35 (1984)
10. A. Richter, B. Straube, W. Pompe: J.Non-Cryst.Sol. 74, 147 (1985)
11. N. Rivier, D.M. Duffy: J.Physique 43, 293 (1982)
12. D.R. Nelson: Phys.Rev. B28, 5515 (1983)
13. B. Lewis: Thin Solid Films 1, 85 (1967)
14. W. Pompe, A. Richter, W. Kreher: unpublihed report ZFW 1985
15. Z. Hashin, S. Shtrikman: J.Mech.Phys.Sol. 11, 127 (1963)
16. W. Kreher: unpublished report ZFW 1985
17. R. Berman: in: The Properties of Diamond (Ed.: J.E. Field) (Academic Press, New York 1979)
18. W. Pompe: Wiss.Ber. ZFW Nr. 23, 157 (1982)
19. W. Pompe, K. Bewilogua, C. Weißmantel, H.-J. Scheibe, A. Richter, H.-D. Bauer: "On the influence of residual stresses and density fluctuations on the crystallization of amorphous carbon films", to be published
20. J.W. Christian: The Theory of Transformations in Metals and Alloys (Pergamon, New York 1965)

Part IV

Evolution Processes

Turing Structures in Cytokinesis, Mitosis and Early Embryogenesis

A. Hunding

Panum Institute, University of Copenhagen, Department of Chemistry 6-2-30, Blegdamsvey 3, DK-2200 Copenhagen N, Danmark

1. Introduction

The problem of spontaneous pattern formation in biology, and the autoregulation of such patterns, is one of the major challenges for theoretical biology. What is to be explained is the branching, or bifurcation, of inhomogeneous (spatially nonuniform) solutions from the initial homogeneous solution of the governing equations. Although this process ultimately should be under control of DNA, it is not immediately obvious how the turning on and off of certain genes could give rise to stable inhomogeneous, well-controlled biological patterns. It was shown by TURING /1/, however, that even simple autocatalytic chemical networks coupled to diffusion may exhibit such spontaneous pattern formation in a predictable and well-controllable way: When parameters such as the size of a cell (or blastula) or an enzyme concentration increase beyond well-defined critical values, the homogeneous solution becomes unstable and prepatterns, i.e. inhomogeneous, stable concentration gradients, emerge spontaneously, without any outside control imposing the symmetry of the prepattern, see Fig. 1.

The emergence of inhomogeneous, stable concentration gradients, apparently from out of nowhere, may be hard to a digest for the uninitiated. Quite a lot of related processes are well established experimentally, however, in numerous branches of science. Examples are the transition of laminar flow of a liquid into vortex generation by increasing the flow, or the Benard instability, in which a liquid heated gently from below develops a highly patterned convection mode by spontaneously creating circulating rolls of liquid. Such phenomena of selforganization in physics, chemistry and biology are mathematically related, and embraced in the new field synergetics /2/.

Most work on reaction-diffusion bifurcations has been confined to one or occasionally to two space dimensions. The present author

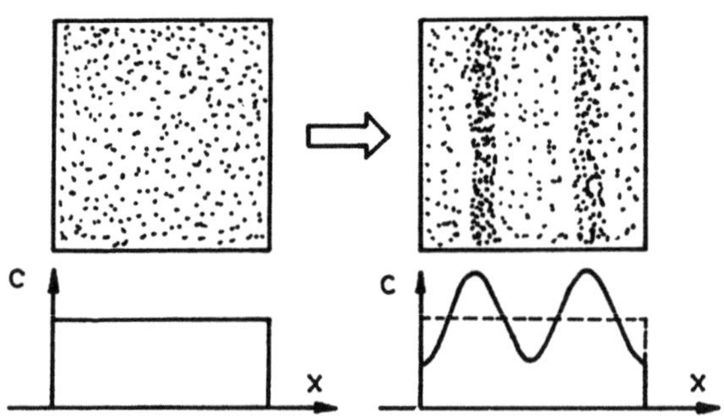

Fig. 1 The homogeneous solution becomes unstable (l.h.s.) and a stable inhomogeneous concentration field originates (r.h.s.)

has extended such numerical studies to three spatial coordinates /3/, and found correspondence between the numerically recorded prepatterns of the three-dimensional sphere and previous experimental work on nuclear division in the radiolarian Aulacantha Scolymantha, and suggested a role of such prepatterns in spatial control in cytokinesis and mitosis /4,5/.

Current models of mitosis are centriolar free models, resting on intrinsic properties of the spindle. The validity of former mitosis models based on the centriole as the prime organizing element has been questioned seriously during the last decades. The evidence of non-centriolar mitosis has accumulated /6/. For example, DIETZ /7/ demonstrated that when the centriolar region is displaced off the spindle of the crane fly, the cells were still perfectly capable of undergoing anaphase and cytokinesis. Recently, laser microbeam destruction of prophase centrioles was accompanied by normal anaphase movements of chromosomes /8/. Indeed, most higher plants undergo mitosis and cytokinesis without centrioles. Among lower eukaryotes, non-centriolar mitosis often occurs. Such facts strongly suggest that the centriole is not needed at the pole during mitosis.

Modern theories on mitosis are thus centriolar-free models, based on intrinsic properties of the spindle.

Recent reviews are given by INOUE /9/ and PICKETT-HEAPS, TIPPIT & PORTER /10/. However, counter experiments to such models are known. It has been shown that anaphase movements of chromosomes are not affected by disruption of interzonal microtubules /11/ a result also found in plant spindles /12/. The related process of cytokinesis has been reviewed for animal cells by RAPPAPORT /13/. He concluded that the asters may induce furrowing in astral overlap regions. Although based on a long series of experiments, which demonstrate that asters are active in determining furrow position and furrow formation, this model also has been criticized. In /14/, ultrastructural evidence goes against aster microtubules crossing at the equatorial cortex at any stage of mitosis relevant to cleavage stimulation. Also, in /15/ destriction by stirring of astral overlap regions did not prevent furrowing. This has led to a revival of competing models such as that of WOLPERT /16/, SCHROEDER /17/ and WHITE & BORISY /18/.

The morphogen prepattern theory, suggested by HUNDING /4/ and discussed here, may be seen as a sort of revival of yet another old idea for cytokinesis, that of a cleavage substance being present in the cell. Although RAPPAPORT /13/ reviews experiments against a pure "cleavage substance" model, some experiments are in favour of the presence of such a substance /19/. This will be discussed below.

2. Prepattern Formation in Reaction-Diffusion Systems

The system under discussion is a chemical reaction-diffusion system in a thermodynamically open system, which may be written in some generality as

$$\frac{\partial C_1}{\partial t} = g_1(C_1,C_2) - \alpha g(C_1,C_2) + D_1 \Delta C_1 , \qquad (1)$$

$$\frac{\partial C_2}{\partial t} = \alpha g(C_1,C_2) - g_2(C_1,C_2) + D_2 \Delta C_2 . \qquad (2)$$

The model includes morphogens with local concentrations C_1 and C_2. Component one is fed (homogeneously) by a source term $g_1(C_1,C_2)$ and degraded to component two by the term $\alpha\, g(C_1,C_2)$, where α is proportional to the concentration of a control enzyme catalyzing this step. Component two is then generated by the same rate, and decomposed according to the term $g_2(C_1,C_2)$. Both substances are free to diffuse within the boundaries of the region, i.e. Fickian diffusion is included in (1) and (2). If the system contains autocatalysis (i.e. Hill-type activation $g(C_1,C_2) = C_1C_2^{\gamma}/(K+C_2^{\gamma})$, where C_2 is a product activator for the decomposition of C_1), the homogeneous steady state to (1)-(2) may become unstable, and stable inhomogeneous concentration profiles may result, such as spontaneous morphogen gradient formation, without any imposing of such a symmetry from the outside. The prepatterns which emerge are closely related to eigenfunctions Ψ_{nlm} of the Laplacian for the region in question, e.g.

$$\Psi_{nlm}(r,\Omega) = j_n(K_{nl}r)Y_{nm}(\Omega) \qquad (3)$$

for the sphere of radius R, with zero flux boundary conditions imposed at r = R. It is possible to show by analytical primary and secondary bifurcation theory, that the bipolar prepattern $j_2(K_{21}r)$ $Y_{20}(\Omega)$ emerges, when enzyme activity α in (1)-(2) is increased beyond a certain critical value α_c (see HUNDING /5/). This has been confirmed by numerical studies (see Fig. 2).

The prepattern theory of mitosis and cytokinesis suggest, that this bipolar prepattern is ideally suited as spatial governor of spindle formation, and positioning and triggering of formation of cleavage furrow or vesicles in animal and plant cytokinesis, respectively. Here the following questions will be addressed: a) Are Turing structures present in cells at all? b) Is the bipolar prepattern, Fig. 2, present during mitosis? c) Are Turing structures involved in (early) embryogenesis; Finally, the possible evolutionary exploitation of Turing structures will be discussed.

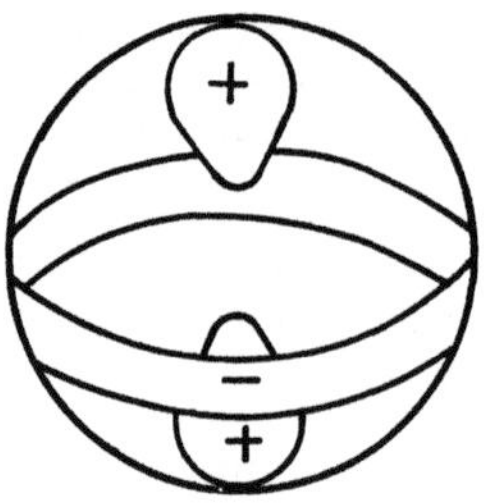

Fig. 2 In analogy with the cosine-function emerging in one dimension, see Fig. 1, in the three-dimensional sphere a pattern j_2Y_{20} emerges with high (+) concentration at two spontaneously arising poles, and low (-) concentration in a ring in the equatorial region. This is called the "mitosis" prepattern.

3. Experiments supporting the prepattern theory

Experimental evidence for Turing structures in cells was claimed to be found in nuclear division of the radiolarian Aulacantha Scolymantha /4/. The more than 1000 chromosomes are confined to highly symmetrical, three-dimensional regions within the nucleoplasm. The emergence of these regions was entirely obscure on the basis of conventional spindle theories, but is nicely explained by Turing structures obtained numerically for the sphere. Especially, a highly symmetrical saddle-shaped chromosome distribution is explainable as the null-region of the recorded pattern $j_1Y_{10}+j_2Y_{22}$, consider Fig. 3.

The question as to the presence of Turing structures in mitosis was also addressed in /4/. Most other theories of the mechanism of mitosis yield forces on the kinetochore-to-pole fibers proportional to fiber length. However, experimental measurements on trivalent

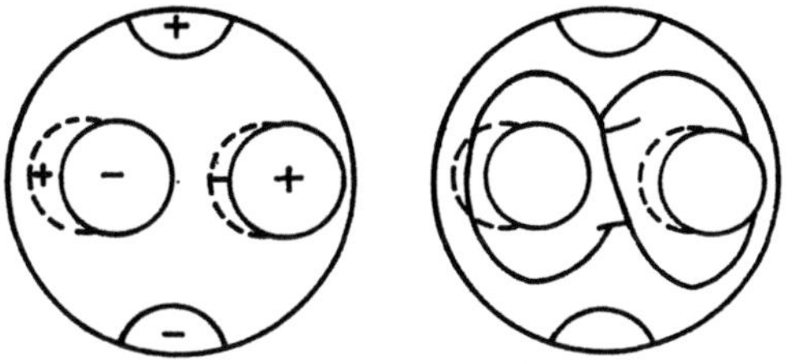

Fig. 3 The null-region of the prepattern $j_1Y_{10}+j_2Y_{22}$ (left) creates a saddle-shaped chromosomal distribution (right), photographed in Aulacantha Scolymantha. Such highly symmetrical distributions are difficult to explain by conventional spindle forces, and may be experimental evidence of Laplacian prepatterns.

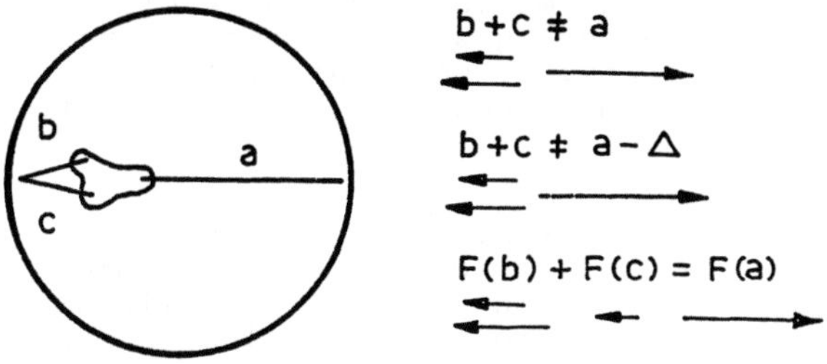

Fig. 4 Explanation see text and Table 1

chromosomes, with two fibers toward one pole, and one long equator-crossing fiber to the other pole, are not consistent with forces acting along the fibers from a potential proportional to the local prepattern concentration. This yields a much better force balance /4,5/.

Turing structures may be involved in the positioning and/or triggering of cytokinesis. The mitosis prepattern, Fig. 2, has a ring of prepattern substance in the equatorial plane. This substance may initiate cytokinesis. A test of this may be carried out by altering the geometry of the cell. Compression of the cell to a flat cylinder yields an "abnormal" form of cytokinesis: direct division to four cells are encountered, two cells contain asters but two do not (Scotts experiment). This is not easily accommodated by most other cytokinesis models, but compression of the bipolar mitosis Turing pattern was shown by HUNDING /19/ to yield a quadripolar prepattern, which supports the observed phenomenon, see Fig. 5.

Table 1 Fiber lengths a, b and c, compare Fig. 4, as measured on resting trivalents (in μ). ep is the equator-to-pole distance. a-(b+c) is consistently much too long. Forces calculated from the prepattern theory F(a)-(F(b)+F(c)), with unmeasured pole-to-surface distance set equal to 10% of ep, are in much better force balance. Equations for F are given in /4/.

a	b	c	ep	a-(b+c)	F(a)-(F(b)+F(c))
14.2	4.7	7.0	12.2	+2.5	-1.5
16.4	3.9	5.9	12.7	+6.6	0.3
15.8	4.6	5.9	12.6	+5.3	-0.3
12.3	2.8	5.3	10.3	+4.2	0.9

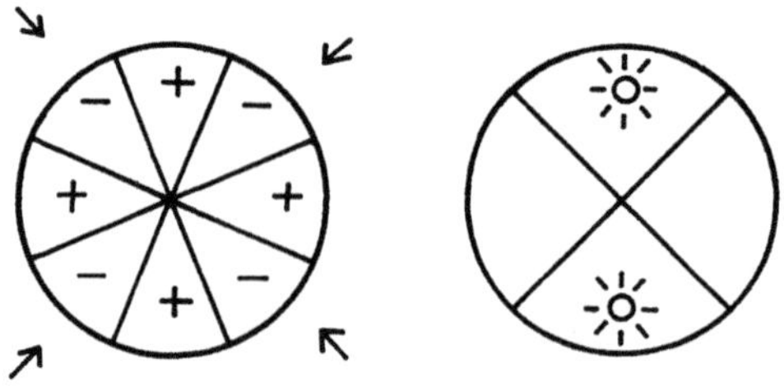

041

Fig. 5 Scott's quadripartition. Highly flattened cells, with axis ratio τ = 5.0, were shown by SCOTT to cleave into four cells, although only two spindle poles are present (right). This can be explained by the prepattern theory, since the emergence of the pattern (041) has four (plus) poles, and four intersecting (minus) regions. A cleavage substance, which used to accumulate in the equatorial ring of the spherical "mitosis" prepattern j_2Y_{20}, Fig. 2, would accumulate in the four regions shown by arrows, in the compressed cell. Quadripartition is thus supported.

Turing structures in early embryogenesis may have been observed recently by experimental recording of spatial distribution of mRNA of the ftz^+gene in Drosophila /20,21/. Simple gradient prepatterns have for a long time been assumed by many biologists, although few link them to Turing structures. In Drosophila however, a seven fold repeat prepattern is recorded experimentally, and this is considered by many to be unexplainable by simple sources or sinks, but in support of Turings mechanism.

From an evolutionary point of view, it is satisfying that the very same principle seems to be capable of governing both cytokinesis, mitosis and early embryogenesis. In /4/ it was suggested that the exploitation of Turing structures for cytokinesis came first, as crude segregation of many identical chromosomes may be achieved by simple cleavage (without mitosis) as seen in some dinoflagellates. Turing structures may also be involved in the distribution of a single pair of daughter chromosomes to each cell, as in bacterial cell division. From this, mitosis linked to microtubuli may have evolved, as the genome was spread to few, different chromosomes.

One objection to Turing structures as spatial governors has been that the pattern does not adapt well to varying sizes of the cell or embryo. This obstacle may be overcome if effective diffusion constants are controlled by binding of the prepattern substances to quasistationary (say, gel-like) components in the cytoplasm. In /19/, models are discussed with simple modification to Turings original equations, which yield size-adaptable prepatterns. Such mechanisms may be involved in control of early embryogenesis, such as blastula cleavage plane orientation, establishment and maintaining of various organization relative to the animal-vegetal axis, like the dorsal-ventral axis, and general specification of positional information in initial embryos less than a millimeter long /22/.

In conclusion, Turing structures may be central to a variety of phenomena in the evolution of life, as an underlying mechanism upon which chemical selforganization, or biocybernetics, rests.

References

1. A.M. Turing: Phil.Trans.Roy.Soc. London B237, 37 (1952)
2. H. Haken: Advanced Synergetics (Springer, Berlin Heidelberg New York 1983)
3. A. Hunding: J.Chem.Phys. 72, 5241 (1980)

4. A. Hunding: J.Theor.Biol. 89, 353 (1981)
5. A. Hunding: in: Evolution of Order and Chaos (Ed.: H. Haken) (Springer, Berlin Heidelberg New York 1982)
6. S.P. Peterson, M.W. Berns: Int.Rev.Cytol. 64, 81 (1980)
7. R. Dietz: Chromosomes Today 1, 161 (1964)
8. M.W. Berns, S.M. Richardson: J.Cell Biol. 75, 977 (1977)
9. S. Inoue: J.Cell Biol. 91, 131 (1981)
10. J.D. Pickett-Heaps, D.H. Tippit, K.R. Porter: Cell 29, 729 (1982)
11. J.A. Snyder, S.I. Vogt: J.Cell Biol. 95, 305a (1982)
12. J. Mole-Bayer, A.S. Bajer: J.Cell Biol. 96, 527 (1983)
13. R. Rappaport: Int.Rev.Cytol. 31, 169 (1971)
14. C.F. Asner, T.E. Schroeder: Expl.Cell Res. 122, 327 (1979)
15. R. Rappaport: J.exp.Zool. 206, 1 (1978)
16. L. Wolpert: Int.Rev.Cytol. 10, 163 (1960)
17. T.E. Schroeder: Expl.Cell Res. 124, 231 (1981)
18. J.G. White, G.G. Borisy: J.theor.Biol. 101, 289 (1983)
19. A. Hunding: J.theor.Biol. 114, 571 (1985)
20. E. Hafen, A. Kuroiwa, W.J. Gehring: Cell 37, 833 (1984)
21. J. Slack: Nature 310, 364 (1984)
22. L. Wolpert: in: Pattern Formation (Eds.: G.M. Malacinski, S.V. Bryant) (Collier MacMilland Publishers, London 1984)

Qualitative Properties of Chains as Models of Growth and Binary Coding

M. Peschel and W. Mende

Akademie der Wissenschaften der DDR, Zentralinstitut für Kybernetik, Rudower Chaussee 9, DDR-1199 Berlin, GDR

1. Introduction

The simplest way to model chains of rate-coupled systems is by Exponential Chains, introduced in /1/. Exponential Chains are composed of modules with

$$dx_i/dt = K_i x_i x_{i+1} \ .$$

For finite chains we have $i = 0,1,\dots,N$ and put $x_{N+1} = 1$. They are suitable to model monotonous growth processes with monotonous logarithmic derivatives of a certain order.

There is an interesting correspondence between Exponential Chains and Taylor Chains with the basic module /2/

$$dx_i/dt = K_i + L_i x_{i+1}$$

defining the Taylor series expansions for $x_0(t)$.

In this paper we study the qualitative properties of Exponential Chains by introduction of so called monotony classes of the solutions on the resolution force given by a finite Exponential Chain. We construct as a tool for the qualitative study a finite automaton with the corresponding monotony classes as states. There is an important relation between the Exponential Chain and a certain binary code for real numbers with the code generator function

$$z = a^{sK\, z'} \quad \text{with } z' \in [0,\infty) \text{ and } s = +1 \text{ or } s = -1$$

These considerations can be essentially generalized by construction of different types of binary codes with general code generator functions

$$z = g(s,K,z') \text{ with } z' \in A \text{ (given interval), } s \text{ binary variable .}$$

This code generator function is a quite natural vehicle to study qualitative properties as convergency, robustness etc. of a corresponding chain concept with the inverse basic module

$$x_i(t) = g(s,K_i,I\, x_{i+1}) \text{ with } I = \int_0^t$$

In this sense the usual dual code of numbers can be used for the analysis of special cases of Taylor Chains.

2. Qualitative Properties of Exponential Chains

For studying the qualitative properties of arbitrary Exponential Chains we first restrict on

$$K_i = \begin{cases} +1 \\ -1 \end{cases} .$$

For the corresponding signal $x_{0,N}(t)$ at the bottom of the chain with all $x_i(0)=1$ we introduce the notation

$$x_{0N}(t) = e(s_0, s_1, \ldots, s_N) \text{ with } s_i = \text{sign } K_i = \begin{cases} +1 \\ -1 \end{cases} .$$

We call these functions coordinate functions of the Exponential Chain of order N+1. Obviously we get by integration the following first- and second-order coordinate functions

$$e(s_0) = e^{s_0 t} , \quad e(s_0, s_1) = e^{(s_0/s_1)\,(s^{s_1 t}-1)} .$$

If we take into account that $\exp(K \int_0^t)$ for $K > 0$ is a positive monotonous and for $K < 0$ a positive antitonous operator, we get the following monotony relationships

$$e(+) > 1 > e(-)$$

$$e(+,+) > e(+) > e(+,-) > 1 > e(-,-) > e(-) > e(-,+)$$

$$e(+,+,+) > e(+,+) > e(+,+,-) > e(+) > e(+,-,-) > e(+,-) > e(+,-,+)$$
$$> 1 >$$
$$e(-,-,+) > e(-,-) > e(-,-,-) > e(-) > e(-,+,-) > e(-,+) > e(-,+,+) .$$

The study of the qualitative properties of Exponential Chains is based on this half-order of the coordinate functions respective of the corresponding binary vectors $(s_0, s_1, \ldots, s_N)$.

This splitting of the lower-order coordinate function into coordinate functions of the next higher order can be transparently represented as shown in Fig. 1. The corresponding linearly ordered set of binary vectors of length l we designate with $\text{or}(W_l)$, and the corresponding reverse order with $\overline{\text{or}}(W_l)$. Obviously the generation rule for the construction of $\text{or}(W_{l+1})$ recursively from $\text{or}(W_l)$ [resp. $\overline{\text{or}}(W_l)$] is given by

$$\text{or}(W_{l+1}) = \begin{cases} + \ \text{or}(W_l) \\ \\ - \ \overline{\text{or}}(W_l) \end{cases} .$$

Using this half-order for binary vectors $W_l = (s_0, s_1, \ldots, s_{l-1})$ we introduce now so-called monotony classes for binary vectors of arbitrary length, namely

```
                      + + +
            + +
                      + + -
      +
                      + - -
            + -
                      + - +

                      - - +
            - -
                      - - -
      -
            - +
                      - + +
```

Fig. 1 Splitting of the lower-order coordinate functions into coordinate functions of the next higher order

$$>W_l = \begin{cases} W_l + \text{or}\ (W) & \text{if} \quad \prod s_i = +1 \\ W_l - \text{or}\ (W) & \text{if} \quad \prod s_i = -1 \end{cases}$$

$$<W_l = \begin{cases} W_l - \overline{\text{or}}\ (W) & \text{if} \quad \prod s_i = +1 \\ W_l + \overline{\text{or}}\ (W) & \text{if} \quad \prod s_i = -1 \end{cases} .$$

Here is W an arbitrary binary word.

Corresponding to the coordinate functions $e(s_o,s_1,\dots,s_{l-1})$ we introduce so-called monotony classes for trajectories by considering one-parametric families of curves in the following way

$$>e(s_o,s_1,\dots,s_{l-1}) = \begin{cases} e(s_o,s_1\dots,s_{l-1},+K_l) & \text{if} \quad \prod s_i = +1 \\ e(s_o,s_1,\dots s_{l-1},-K_l) & \text{if} \quad \prod s_i = -1 \end{cases}$$

$$<e(s_o,s_1,\dots,s_{l-1}) = \begin{cases} e(s_o,s_1,\dots,s_{l-1},-K_l) & \text{if} \quad \prod s_i = +1 \\ e(s_o,s_1,\dots,s_{l-1},+K_l) & \text{if} \quad \prod s_i = -1 \end{cases} ;$$

$>e(s_o,s_1,\dots,s_{l-1})$ is the upper and $<e(s_o,s_1,\dots,s_{l-1})$ the lower monotony class for $e(s_o,s_1,\dots,s_{l-1})$. $e(s_o,s_1,\dots,s_{l-1},s_lK_l)$ is the solution $x_{ol}(t)$ of a chain with the parameters $s_o,s_1,\dots,s_{l-1},s_lK_l$ $(K_l > 0)$.

Obviously these classes form bands of curves completely covered by the corresponding curve families parametrized by the parameter $K_l > 0$. It can be seen easily that all these monotony classes together are a decomposition of the whole positive quadrant into disjoint bands.

Without the above-mentioned parameter restriction, we can in the same way consider coordinate functions $e(s_oK_o,s_1,K_1,\dots,s_{l-1}K_{l-1})$ and the corresponding upper and lower monotony classes $>e(s_oK_o,s_1K_1,\dots,s_{l-1}K_{l-1})$ and $<e(s_oK_o,s_1K_1,\dots,s_{l-1}K_{l-1})$.

The monotony classes thus constructed for a given l (resolution power) can be considered as classes (of trajectories) in the sense of the theory of classification. But at the same time we are going to consider the monotony classes as states of a finite automaton.

We will now study how the state changes if we adjoin another basic module either at the top of a given chain or at the bottom with fixed value of l.

Let us first adjoin another base module at the top of the Exponential Chain, meaning that we consider now the chain with the parameter

$$s_oK_o, s_1K_1, \ldots, s_{l-1}K_{l-1}, s\ K \ .$$

From the considerations presented above, we know that the corresponding coordinate function

$$e(s_oK_o, s_1K_1, \ldots, s_{l-1}K_{l-1}, s\ K)$$

either belongs to the state

$$> e(s_oK_o, s_1K_1, \ldots, s_{l-1}K_{l-1}) \text{ or } < e(s_oK_o, s_1K_1, \ldots, s_{l-1}K_{l-1})$$

depending on the sign of s:

$e(s_oK_o, s_1K_1, \ldots, s_{l-1}K_{l-1}, sK)$ belongs to $> e(s_oK_o, s_1K_1, \ldots, s_{l-1}K_{l-1})$ for $\Pi s_i\ s = +1$ and to $< e(s_oK_o, s_1K_1, \ldots, s_{l-1}K_{l-1})$ for $\Pi s_i\ s = -1$.

Thus in any case an unique state of our automaton is specified independent of how many additional modules we adjoin at the top level of the Exponential Chain for a fixed value of l.

Therefore we get the following property:

If we classify the possible outputs $x_o(t)$ of a chain which begins with the parameter sequence $s_oK_o, s_1K_1, \ldots, s_{l-1}K_{l-1}$ with resolution force l (that means if we use 2^l monotony classes), the behavioural type of all longer chains is uniquely determined by the initial chain

$$s_oK_o, s_1K_1, \ldots, s_{l-1}K_{l-1} \ .$$

Now we shall adjoin another basic module at the bottom level of the Exponential Chain with the parameters $s_oK_o, s_1K_1, \ldots, s_{l-1}K_{l-1}$, that means we consider the following chain specified by the parameter sequence

$$sk,\ s_oK_o, \ldots, s_{l-1}K_{l-1} \ .$$

The signal $x_o(t)$ generated at the bottom of this chain can be interpreted as if we had adjoined the basic module with the parameters $s_{l-1}K_{l-1}$ at the top level of chain specified by the parameters

$$sk,\ s_oK_o, \ldots, s_{l-2}K_{l-2} \ .$$

We meet now the difficulty that by this manipulation all coordinate functions are quantitatively changed.

But it is important to mention that qualitatively the corresponding state transition depends only on the signs s_i.

Therefore we look only on the state transition for the case with all $K_i=1$. In this case the following transition rule can be easily derived:

$$s \gtrless e(s_0,s_1,\ldots,s_{l-1}) = \begin{cases} > e(s,s_0,\ldots,s_{l-2}) & \text{for } \prod s_i \quad s = 1 \\ < e(s,s_0,\ldots,s_{l-2}) & \text{for } \prod s_i \quad s = -1 . \end{cases}$$

Here s designates the operator corresponding to the chain extension by adjoining a basic module with the sign s at the bottom.

Most important is the fact that we really get a state-transition and that none of the states is split up. The result can be illustrated by graphs corresponding to our classifying automata. Fig. 2 shows the graphs for the resolution powers l = 0,1,2 and 3.

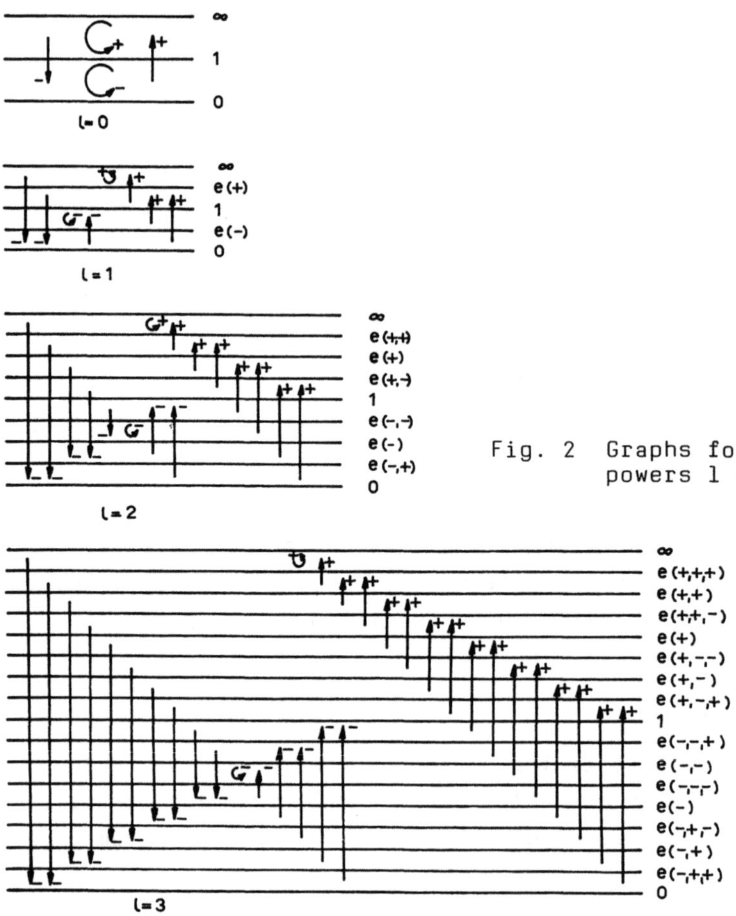

Fig. 2 Graphs for the resolution powers l = 0,1,2 and 3

From these graphs we recognize immediately the remarkable role played by hyperbolic growth (as a monotonous class attractor) and hyperbolic decay (as a focus-like class attractor).

3. A Binary Code Corresponding to the Exponential Chains

We are interested to code the coordinate functions $e(s_0, s_1, \ldots, s_{l-1})$ in such a way that the order relation between the corresponding binary vectors $s_0, s_1, \ldots, s_{l-1}$ is reflected in the value of the code.

This is obviously achieved, if we introduce for a given a, with $a > 1$ the following Exponential Code

$$z = a^{s_0 a^{s_1 a^{\cdots s_{l-2} a^{s_{l-1}}}}} .$$

This code leads us to study in more detail the corresponding code mapping determined by the following code generator function

$$z = a^{s\, z'} .$$

This function, defined on the interval $[0, \infty)$, maps for s=+1 the interval $[0, \infty)$ on $[1, \infty)$ and for s=-1 on $[1, 0)$, presenting thus a dichotomy of the basic interval $[0, \infty)$ of the code.

If we iterate this code mapping a finite number of times, we produce apparently a dichotomy tree with smaller and smaller intervals contracting to length zero if the number of iterations approaches infinity.

This is a model for the Weierstrass convergency holding, as we have seen above, for the monotony classes of the solutions of an Exponential Chain with increasing length.

Really the contraction property of the partition of the basic interval by iterative application of the code generator function causes the corresponding Weierstrass convergency of the corresponding bands of trajectory because of the preserved order relationship between the code numbers and the corresponding coordinate functions.

For the analysis of the properties of the code, also the inverse code generator function is of importance, which by iteration produces the binary words $s_0, s_1, \ldots$ of the code to a given real number $z \in [0, \infty)$.

In our case, we get for the inverse code generator function

$$z' = \mathrm{abs}(\ln z)/\ln a = \begin{cases} \ln z/\ln a & \text{for } z \in [1, \infty) \quad s=+1 \\ -\ln Z/\ln a & \text{for } z \in (0, 1] \quad s=-1 \end{cases} .$$

Fig. 3 shows the graph of the inverse code generator function. In a certain pragmatic sense the code generator function corresponds to

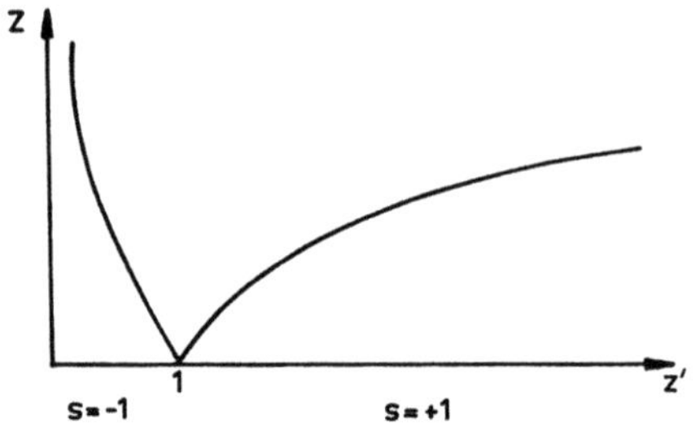

Fig. 3 Graph of the inverse code generator function

integration (in analogy to the exponential integrator), while the inverse code generator function corresponds to differentiation (which can raise chaotic trajectories). In the following section we elaborate a generalization of the correspondence between chains of basic modules and corresponding codes with the tendency to use the codes as a vehicle for the study of the qualitative properties of the dynamic chains.

4. Dynamical Chains and Binary Codes

A binary code is uniquely determined by a code generator function

$$z = g(s,z')$$

obeying the following conditions:

- s is a binary variable with the values s_1, s_2 (arbitrary but fixed)
- the functions $g(s_i z')$ are monotonous functions both defined on a given basic interval J and mapping the interval on intervals J_i, i=1,2 which define a partition of J, namely

  $J_1 \vee J_2 = J$ and $J_1 \wedge J_2 = 0$ (up to at most one point)

If these conditions are fulfilled, we get the following results immediately:

- Depending on character of the monotonous mappings $g(s_i, z')$ we have to distinguish between four types of corresponding partitions of the basis interval
  g_i may be monotonous or antitonous, i=1,2
  Fig. 4 shows these four combinations of monotonous respective antitonous mappings of the basis interval.
- The iterative application of g with variable binary values s_i generates obviously for the iteration number l a partition of the basic interval into 2^l intervals, each of them contracting to a point as l approaches infinity (Weierstrass convergency).

Together with the code generation function g(s,z'), also the corresponding inverse function is of great interest, namely for gene-

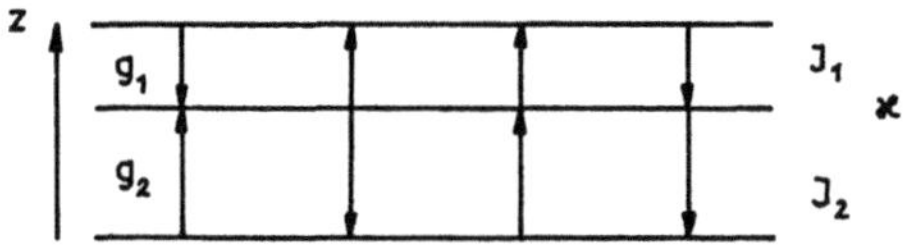

Fig. 4 Four combinations of monotonous respective antitonous mappings of the basis interval

ration of the binary vector $s_0, s_1, \ldots, s_{l-1}, \ldots$ to a given real number $z \in J$, that means the corresponding code word. This function is

$$z' = \begin{cases} g^{-1}(s_1, z) & \text{for } z \in J_1 \\ g^{-1}(s_2, z) & \text{for } z \in J_2 \end{cases} .$$

Using the pragmatic idea that in some sense the code generator function is something like integration, while the inverse code generator function is something like differentiation, we now assign the code generator function a corresponding dynamic chain, namely

$$z = g(s,z') \qquad x_i(t) = g(s_i, I\, x_{i+1}) \quad \text{with } I = \int_0^t \qquad \text{type I*}$$

which can be easily transformed into a differential equation, and we now assign the inverse code generator function a corresponding dynamic chain, namely

$$z' = g^{-1}(s,z) \qquad x_{i+1}(t) = g(s_i, dx_i/dt) \ . \qquad \text{type II*}$$

Because of the contraction property of the partition of J by iterative application of $g(s,z')$, we have to expect the corresponding Weierstrass convergence of the bottom signal $x_0(t)$ of the chain of type I with increasing chain length, while because of the properties of the inverse code function to generate chaotic trajectories we can possibly get from chains of type II a systematical possibility to gain chaotic trajectories.

Some Examples of our Concept for Constructing Dynamical Chains

1. Linear Codes

We study linear codes under the following conditions

$$J = [0,1] \quad J_1 = [\varkappa,1] \quad J_2 = [0,\varkappa] \quad \text{with arbitrary } \varkappa,\ 0 < \varkappa < 1,$$
$$s_1=1,\ s_2=0 \ .$$

For the four cases we get quite simply the following results

- g_1 antitonous, g_2 monotonous
  $g = \varkappa z' + s(1-z')$

* If $I\, x_{i+1}$ or dx_i/dt are leaving the definition interval of the mapping g, suitable nonlinear functions should be applied on these variables.

- g_1 monotonous, g_2 antitonous
  $g = \varkappa(1-z') + s\,z'$
- g_1 monotonous, g_2 monotonous
  $g = \varkappa z' + s(\varkappa + z' - 2\varkappa z')$. This is for $\varkappa = 1/2$ the usual dual code.
- g_1 antitonous, g_2 antitonous
  $g = \varkappa(1-z') + s(1 - z' - \varkappa + 2\varkappa z')$

The corresponding dynamic chains are:

type I: $dx_i/dt = (\varkappa_i - s_i)\, x_{i+1}$

$dx_i/dt = (s_i - \varkappa_i)\, x_{i+1}$

$dx_i/dt = (\varkappa_i + s_i - 2\varkappa_i s_i)\, x_{i+1}$

$dx_i/dt = (2\varkappa_i s_i - \varkappa_i - s_i)\, x_{i+1}$.

These are all special cases of (convergent) Taylor Chains. That means there is an interesting connection between the usual dual code and the Taylor series expansion.

type II: $dx_i/dt = (x_{i+1} - s_i)/(\varkappa_i - s_i)$

$dx_i/dt = (x_{i+1} - \varkappa_i)/(s_i - \varkappa_i)$

$dx_i/dt = (x_{i+1} - \varkappa_i)/(\varkappa_i + s_i - 2\varkappa_i s_i)$

$dx_i/dt = (x_{i+1} - s_i)/(2\varkappa_i s_i - \varkappa_i - s_i)$.

2. Exponential Codes

We study exponential codes under the following conditions

$J = [0,\infty) \quad J_1 = [\varkappa,\infty) \quad J_2 = (0,\varkappa]$ with arbitrary $\varkappa$, $0 < \varkappa < 1$

$s_1 = +1, \; s_2 = -1$.

For the four cases we get quite simply the following results

- g_1 antitonous, g_2 monotonous
  $g = \varkappa \exp(-a/z') + (s+1)/2.\ (\varkappa - \varkappa e^{-a/z'} + K\, e^{a/z'} - K)$ $\quad (a,k > 0)$
- g_1 monotonous, g_2 antitonous
  $g = \varkappa \exp(s\,a\,z')$ with $a > 0$ arbitrary
- g_1 monotonous, g_2 monotonous
  $g = \varkappa(1 - \exp(-az')) + \varkappa(s+1)/2.\ (\exp(az') + \exp(-az') - 1)$
- g_1 antitonous, g_2 antitonous
  $g = \varkappa \exp(-az') + (s+1)/2.\ (\varkappa - K + K\, e^{a/z'} - \varkappa e^{-a\,z'})$.

To derive the corresponding dynamic chains is quite straightforwardly possible at least formally.

We give here only the result for the second case g_1 monotonous, g_2 monotonous

type I: $dx_i/dt = s_i a_i\; x_i x_{i+1}$ Exponential Chain

type II: $dx_i/dt = \dfrac{1}{a_i s_i}\ (\ln x_{i+1} - \ln \varkappa_i)$.

Remark

In an analogue way we can introduce codes with a given finite number of states

$s_1, s_2, \ldots, s_k$

with corresponding monotonous or antitonous functions $g(s_i, z')$ spanning a corresponding code generator function respective the inverse function. In this case $g(s,z')$ is the interpolation polynomial of order k-1 assuming at the points $s=s_i$ the values $g=g_i=g(s_i,z')$. The construction of two types of dynamical chains from the corresponding basic modules can be repeated as was done above. The famous Feigenbaum sequence

$$z' = K\, z(1-z) \quad \text{with } z \in [0,1]$$

can be interpreted as an inverse code function, the corresponding code generator function is then

$$z = \frac{1}{2}\left[1 + s \text{ sqrt } (1-4\, z'/K)\right] \quad ,$$

mapping monotonously for K=4 J= [0,1] on [0,1/2] and antitonously [0,1] on [1/2,1].

References

1. M. Peschel, W. Mende: Leben wir in einer Volterra-Welt? (Akademie-Verlag, Berlin 1983)
   M. Peschel, W. Mende: The Predator-Prey Model: Do we live in a Volterra World? (Second extended edition by Akademie-Verlag and Springer 1986)
2. W. Mende, M. Peschel: Power-Product Structure-Building Phenomena in Systems with Forces. in: Chaos and Order in Nature (Ed.: H. Haken) (Springer, Berlin Heidelberg New York 1981)
3. W. Mende, M. Peschel: Problems of Fuzzy Modelling, Control and Forecasting of Time-Series and some Aspects of Evolution. IFAC Symposium on Control Mechanisms in Bio- and Ecosystems, Plenary Paper, Leipzig 1977

A Study Program of Chaotic Dynamics Applied to Information Processing

J.S. Nicolis

University of Patras, Department of Electrical Engineering, Patras, Greece

1. Statement of the Problem

In earlier papers (NICOLIS 1982, 1983, 1984, 1985) we essentially defended the thesis that chaos may provide the connection between information theory and physics, the main issue being that information is a physical entity and as such it should not be treated in absentia of its dynamical (hardware) substratum.

How can we explain on the basis of known dynamical laws and processes of physics the communication transaction taking place between two (hierarchical) systems, i.e. not only the energetic exchanges but also the information ones? How can we dynamically implement the generation, stable storage, reliable retrieval and dissipation of information? In other words: What are the physical (dynamical) agents mediating symbolic interactions?

In the first place we have to clarify what we mean when we assert that two systems "communicate". For a physicist, information is admittedly an almost "metaphysical" word. Indeed in the current parlance of the mathematical theory of communication (invented by Claude SHANNON 37 years ago and not improved considerably ever since) information processing is to be considered as a sequence of (one-to-one) mappings from (the members of) a set belonging to a (multidimensional) abstract signal-space onto (the members of) another set belonging to an abstract space of a different dimensionality; the "solution" of the problem of information transfer from set I to set II or more concretely from place A at time τ to place B at time $\tau + \Delta\tau$ boils down essentially to a variational principle which tries to find the optimum mapping (or code) in order to compromise two conflicting requirements: speed of transfer and reliability or degree of errorless reproduction of an abstract pattern of digits sent from I and received by II: in such a formulation all physics is left out.

Such a mentality of treating communication processes serves admirably the whole group of applications associated with typical communication engineering design.

People grow restless, however, when the above formalism is applied to fields such as biology, cognitive psychology and in general to interactions between "self"-organized systems (for a definition see below). The reason is that in such cases complete decoupling between physical and symbolic interactions (or between hardware and software) is no more possible or meaningful, and people start asking questions about the goings on at the interface between structure and function.

In such cases, information processing is associated:

a) With cascades of mapping - transcriptions and/or translations - (not necessarily one-to-one but also multi-to-few, i.e. hierarchical) from one physical entity onto another. For example, in the generation of speech a pattern residing in a sequence of nerve pulses incident upon the vocal chords is mapped onto an air sound

wave. This sound wave is intercepted by a microphone which maps the pattern onto an electromagnetic wave, etc.
b) With low-energy triggers giving rise to high-energy effects (when informational channels intersect effector systems) or, to put it another way, with extreme sensitivity to initial conditions.

There is physical justification (see e.g. /5-7/) for the hierarchical or non one-to-one mapping between "object" and "image", that is the fact that physical perception is ambiguous. There are three distinct reasons for this:

a) The loss of evanescent waves: half of the encoded information in the spherical wave scattered from the object is irretrievably lost due to the impossibility of designing receptors for both propagating and evanescent waves, into which a spherical wave is analyzed.
b) The existence of fluctuations around the average number of photons carried by each normal mode or electromagnetic degree of freedom.
c) The finite aperture of the perceiving instrument: This finiteness has two results: it distorts like a spatial filter a portion of propagating components of the impinging spherical wave, and second it attributes a finite number of degrees of freedom on the receiving image. The ambiguity here is due to the fact that the spherical wave which mediates originally the information transfer from the transmitter to the receiver processes an infinite number of degrees of freedom.

At first glance, it seems that there exist irreconcilable differences between soft- and hardware. Specifically the difficulty lies on the fact that at the hardware (physical) levels, processes follow laws with universal validity, whereas at the software (symbolic) levels processes follow rules, whose relativity and flexibility are manifested even at the level of a writer struggling for "le mot juste".

What is the difference between laws and rules? Can we "reduce" the second to the first? Our conjecture at this point is that the issue involves essentially different time scales. If one allows a deterministic physical law to operate long enough the result may be chaotic. However the time scale is quite different say for gravitational turbulence, "cardiac" turbulence (fibrillation), hydrodynamic turbulence or symbolic turbulence (language?) to develop.

The time scale beyond which chaos emerges is in fact dependent not only on the intrinsic nonlinearities and the stability properties of the system involved but also on extrinsic reasons (initial and boundary conditions). Depending on the case, this time scale may be large ($\sim 10^{15}$ years) or small ($\sim$a few minutes).

The above theorizing aims at bridging the (colossal) difference in "behaviour" between say 1) a physical body which when faced with the task of moving from place A to place B follows simply and "compulsively" Maupertuis variational principle and 2) an animal which "chooses" the specific pathway anew every time. In this second case the animal follows a rule or algorithm which may spontaneously modify itself depending on slight changes in the initial conditions and the parameters of the situation. We claim in short that symbolic processes are chaotic manifestations of collective properties of physical systems - at a higher hierarchical level; "Decision making" and "free will" are deterministic since they are harnessed by natural laws but unpredictable since they are chaotic; the same thing could be claimed for music and language.

Be that as it may, we shall argue that the most important aim of the communication transaction is not just copying but simulation: this essentially refers to a process of modelling a physical system by another one with the obvious goal of predicting and controlling it. We say that a physical system I (see Fig. 1) simulates another (II)

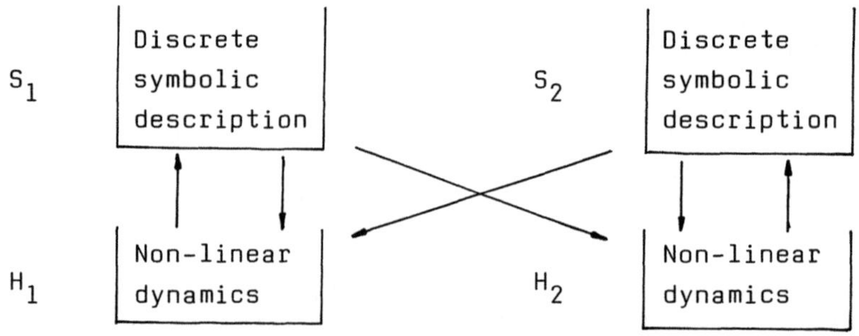

Fig. 1 Communication between two hierarchical systems possessing one Hardware and one Software level

when it is capable, out of a finite length time series it receives, of constructing minimal length algorithms or compressed descriptions of II, which algorithms when introduced as inputs (programs) in a finite-state machine will allow it to furnish as an output the behaviour of system II for all times future and past.

Since the prerequisite for simulation is the compressibility of the received time series from the partner-system (or the environment), we, in this paper, indulge in a new dynamical "paradigm" namely dissipative chaos, which through the gadgetry of "strange attractors" promises in a single stroke to do this : to carry out very complex functional repertoires with very simple "hardware" and to behave as a versatile (albeit somewhat uncontrollable) information processor. The new paradigm comes forth if one realizes that information is not only produced via cascading bifurcations giving rise to broken symmetry, but also via successive iterations giving rise to ever-increasing resolution. Beyond a certain resolution interval, microscopic fluctuations are no longer smeared out; they get amplified and pass from lower to higher hierarchical levels. It is these fluctuations which essentially account for new information generated by an evolving hierarchical system.

In short, the purpose here is to develop an evolutionary approach for the basic problem of generation, storage and dissipation of information in physical and biological systems via dynamical processes. These processes are associated with branching-offs or cascading bifurcations giving rise to broken symmetry as well as chaotic dissipative dynamics giving rise to decreasing and increasing resolution. From an attractor simulating some aspects of a cognitive system (or a "processor") there exist two basic requirements: Big storage capacity and good compressibility. Conventional singularities such as stable steady states and stable limit cycles, possessing dimensionality zero and one, respectively, are very poor as information storage units but ideal as information compression devices.

Strange attractors, on the other hand, via an harmonious combination of expanding and converging trajectories in state space ("positive" λ_+, and "negative" λ_- Ljapunov exponents, respectively) can in principle comply with both requirements: They may possess a considerable information dimensionality - and this idiom makes them suitable for dynamical information storage; on the other hand, being "attractors", ($|\lambda_-| > \lambda_+$) they can serve at the same time as information compressors. (See following Sections for detail.)
We are going to see that (dissipative) chaotic dynamics offers an alternative and perhaps more elegant way towards the formation of hierarchy and classification: Whereas the time-honored statistical-mechanical methods lead to hierarchial structures through the collective properties of a Master equation (e.g. moments and cross-moments of the probability density function describing the dynamics

at the "microscopic" level) dissipative chaotic systems accomplish the same thing differently. They do it by reducing the number of degrees of freedom in state space - where the dynamical system starting from a set of initial conditions of dimensionality N sooner or later locks-in at a compact subset of state space (the "attractor") - of much lower dimensionality.

For system I to be able, in the above sense, to simulate or model system II (and vice versa), a number of sine qua non prerequisites are demanded.

In the first place, the system must be hierarchical, i.e. possess at least a hardware (energetic-structural) level and a software (cognitive symbolic, functional) level. Information, in the form of discrete time series emanating from the S-level of system II, impinges at level H of system I, where a set of convolutions (cross-correlations) takes place among the time series received and the intrinsic dynamics at level H is chaotic. The result of such convolutions is the emergence of collective properties possessing far fewer degrees of freedom than S_2 or H_1. These collective properties are now transported at the level S_1.

The communication or mutual simulation between the two systems may go on, therefore, as the evolution in time of a set of pairs of coupled non-linear integral-difference equations (each associated with a given value of ξ and ϱ).

$$\left.\begin{aligned} S_1(n+1) &= \lambda S_1(n) + \langle H_1^{\xi}(n) S_2^{\varrho}(n) \rangle \\ S_2(n+1) &= \lambda S_2(n) + \langle H_2^{\xi}(n) S_1^{\varrho}(n) \rangle \end{aligned}\right\} \quad \lambda > 1 , \qquad (1)$$

$H_i(n+1) = f_i(H_i(n))$ or $\dot{H}_i = f_i(H)$ being the intrinsic dynamics at the hardware level (s).

Each pair stands in (1) for one statistical moment. For each pair of the above equations the process clearly amounts to a cascading iteration on S_1 and S_2. S_1, S_2 stand for the discrete (in general multidimensional) variables at the software level(s) H_1, H_2 stand for the (discrete or continuous) variables at the hardware levels and n is discrete time; ξ and ϱ are positive integers. Finally $\langle \cdot \rangle$ stand for time averaging.

The harmonious co-evolution of (dynamical) bare variables and their collective properties in time is a demand for any linguistic scheme which includes semantic connotations, i.e. context and meaning: This ongoing iterative interaction (Fig. 1) between a collective property (a "set") with each of its own members and vice versa creates the "context" and at the same time is responsible for the phenomenon of self-reference. Due to a difference in the time scales at the H and S levels we can say that the S levels may exercise history-depending, feedforward control on the H levels. Indeed the integral component brings forth "memory" effects.

We have to emphasize that the above model is quite different in philosophy and physics from the usual "Artificial intelligence" (A.I.) paradigm. A.I. is local theories dealing essentially with logical circuity; in them deductive processes take place where the outcomes are implicitly included in the initial premises. Chaotic processes on the other hand are more akin to inductive reasoning; most of the information is not contained in the initial conditions but is produced "en route" via branching-offs of the dynamical flow itself.

Hierarchy - like self-organization (or the ability of a system to compress environmental time series and construct minimal-length algorithms for the simulation of the environment involved) - is an emergent property of nonlinear dynamical systems, evolving under a discontinuous interplay among the basic (four) types of physical interactions. It is through a study of such an interplay that one can historically understand the ability to communicate as an evolutionary, emergent property of complex large-scale systems (something which runs against the traditional worldview dictated by a static universe running down to its "thermal death"). That is to say, in an expanding universe the interplay of the physical interactions gives rise in a concomitant way to entropy production, progressive differentiation, increasing complexity, and increasing organization. The cosmological, thermodynamic, electromagnetic and signal-processing "arrows of time" all point in the same direction: the direction of increasing differentiation with time, and the emergence of cognitive systems which not only can simulate the external environment but, in a certain imcomplete way, even parts of themselves.

Let us now deliberate on the concepts of hierarchy complexity and organization.

A. Hierarchy

By a hierarchical system we mean, in general, an ensemble of interacting parts that is composed of (and is analysable or decomposable into) successively nested sets of interacting subunits.

Each set of interacting components - forming a distinct hierarchical level - affords a specific state-space description with variables and parameters pertaining to that particular level. The interacting variables (and/or parameters) at a higher hierarchical level are "collective properties" (statistical moments or convolutions) of the dynamics going on at the level just underneath. Therefore as we move to a higher level, we (usually but not always - see last section) witness a tremendous reduction in the number of degrees of freedom. The higher level receives selective information from "below" and in turn it exercises on the dynamics of the lower level (efferent) feedforward control commands - which harness the further dynamical deliberations on that lower level.

Although two systems do not have to be hierarchically predisposed in order to exchange energy, in order to enter in a communication transaction they certainly do. (We need one hierarchical level to perform a cross correlation and another higher one in order to "cognize"it.)

B. Complexity

A system may be complex either on a structural or a functional level. Structural complexity increases with the number of interacting subunits; the percentage of mutual connectedness among them, in dyadic or plural fashion; and the variance of the probability density function of the strengths of interactions between the individual subunits.

On a functional level, complexity increases with the minimum length of the (most compressed) algorithm from which one can retrieve the full behaviour of the system. In this sense, as we are going to see in the paper, the functional complexity of a system increases with the length of its "phylogenetics tail", namely its evolutionary history marked by a discrete sequence of bifurcation episodes. In that sense, then, a virus or an orange is a more complex system, functionally speaking, than a big star, since in order to understand the function of each one of them you already need as a prerequisite the existence of a supernova. But let us begin from the beginning:

Consider a macroscopic physical system at its lowest hierarchical level, namely the level of equilibrium thermal (molecular) mixing. At that particular state the system possesses the maximum number of degrees of freedom or the fullest possible symmetry - exactly equal to the symmetry of the microscopic laws of motion of its constituent interacting parts. The symmetry is then maximum, the number of degrees of freedom likewise, and the complexity is minimum since the number of bits required to specify a system at thermodynamic equilibrium is the shortest possible: To reproduce such a system one may start from almost any initial condition.

Imagine now that the system comes in contact with the environment; it may again settle to equilibrium when new conditions for minimization of its free energy are met; or it may happen that as a consequence of internal and/or external microperturbations (which are inevitable due to the enormous number of the constituent components of the system and its environment) the one and only stable steady state of equilibrium may become deleted. Beyond a certain critical distance from equilibrium (which the system franchises triggered by fluctuations) at least two new, alternative stationary configurations (containing their own symmetrical branches) emerge. Suppose the system settles in one of them by a more or less random "choice". At this metastable state the system now forms, say, a spatial static pattern (i.e. a crystal) which, although it may appear quite symmetrical, has a number of degrees of freedom far below the one corresponding to the symmetry of the laws of motion of its constituent molecules. Nevertheless, we do not yet speak of "broken symmetry" but rather of "repealed" symmetry /8/ since the probability of "tunneling" to the symmetrical metastable state remains finite (although it rapidly decreases with the scale of the system).

If, however, by coupling the system with an external asymmetric field we perturb the bifurcation diagram so that the alternative mirror-symmetrical configurations get repressed /9/, we do witness a genuine broken symmetry and the asymmetrical configuration at the corresponding state may be stable for a wide range of the control parameters. We say that the complexity of the system has increased, meaning that the length of instructions or the minimum amount of bits an external observer requires for the replication of the system has increased. We can imagine a succession (cascade) of such symmetry breaking, bifurcation "episodes" stimulated by the shifting of the numerical values of the control parameters beyond critical thresholds; these broken symmetries increase the complexity of the system by sequentially reducing the number of degrees of freedom of the new pattern (s) (via triggering long-range spatial-temporal correlations). The greater the reduction of symmetry i.e. the greater the number of constraints a pattern incorporates, the greater the length of the minimum program necessary for the "complete" description of the system - which would lead to its replication.

So we may consider that each pattern (at a hardware level) is associated with a number of digits (at a software level) - we could classify them as belonging to two categories "pre"-digits and "post"-digits. Pre-digits compose the set of the successive critical (threshold) numerical values of the control parameters, which triggered the corresponding sequence of bifurcations whose final product of broken symmetry is the pattern under consideration. The more abstract the pattern is, the longer the sequence of pre-digits needed to mark (and preserve) its "dotted" or "punctuated" evolutionary history. We may say that the pre-digit set is the evolutionary "zip-code" of the pattern. If the set of pre-digits constitutes the "address" of the pattern, the set of post-digits constitutes its "ID card" or rather its "fingerprints": the set of post-digits contains the set of the amplitudes of the "order parameters" /10/ or the dominant eigenfunctions which emerge after an instability. These eigenfunctions form the collective or macroscopic properties of the pattern concerned.

If a pattern dynamically produced as above is now described or coded using a string of $\{0,1\}$ symbols, the length of the shortest such string can be deduced as follows: Let p be the a priori probability of 0's and 1-p that of the 1's in a fair coin tossing realization of length N digits (bits). There are $\binom{N}{pN}$ such sequences, or

$$\frac{N!}{[(pN)!][N(1-p)!]} .$$

Hence $\log_2\binom{N}{pN}$ bits suffice to uniquely specify such a sequence, or

$$\frac{\log_2\binom{N}{pN}}{N} \frac{\text{bits}}{\text{digit}}$$ using Stirling's formula

$$\xi! \approx \sqrt{2\pi\xi}\ \xi^{\xi} e^{-\xi} \quad \xi \gg 0$$ and applying it above, we get the expression

$$S(p) = -p \log_2 p - (1-p) \log_2 (1-p) \frac{\text{bits}}{\text{digit}}$$

for uniquely specifying a N sequence of 0 and 1 with pN symbols 0 and (1-p)N symbols 1 (as $N \rightarrow \infty$).

However, this is not all: To reconstruct a given finite length N sequence of 0's and 1's where p is the a priori probability of 0's, one needs to specify the number of zeros which for N finite is not exactly known, that is one needs an extra amount of information $\log_2 (n+1)$ bits long*. So the total complexity of the pattern is

$$N' \sim \log_2(N+1)+\log_2\binom{N}{pN} \cong \log_2 N + NS(p) \quad \text{for } N \gg 1 \text{ or just}$$

$$S(p) \frac{\text{bits}}{\text{digit}} \quad \text{as } N \rightarrow \infty .$$

CHAITIN /11/ showed clearly that the overwhelming majority of the above sequences are incompressible, i.e. random.

Indeed, consider the number of sequences of given length N which are compressible up to K bits, i.e. $N' \leqslant N-K$. This number equals

$$\Sigma = 2^1 + 2^2 + \ldots + 2^{N-K-1} = 2^{N-K} - 2 .$$

The percentage of compressible sequences is thus

$$n = \frac{\Sigma}{2^N} \sim 2^{-K} \quad (N \gg 1) .$$

So, for K = 10 for example, only one in a thousand sequences can be specified by an algorithm 10 bits shorter that the sequence itself.

This fact prompts us to suggest a dynamical gadgetry for the memory of our receiver. Instead of storing in the memory bank of this receiver the full repertoire of the transmitter messages (something quite impractical for large N), one rather implements at the hardware level of the decoding system an algorithm, i.e. a non-linear recurrent relationship (map) which upon request, i.e. after being triggered from the transmitter of the partner-system, puts forth "here and now" a whole set of aperiodic strings of digits - whose "grammar" is tempered by the shape of the map. Most of them will be pseudorandom sequences, and hence the probability of matching any incoming "packet" of given length N will go to one, asymptotically with time.

C. Organization

A cognitive device capable of compressing and simulating a physical phenomenon possesses a more or less advanced degree of "self"-organization.

---

* N+1 digits are needed instead of N, just in order to make the difference between N zeros and N ones.

In general, self-organization is associated with the spontaneous emergence of long-range spatial and/or temporal coherence among the variables of the (organized) system. Nevertheless, the system may be quite "innocent" of this selforganization, namely it may not use it further or may not know how to use it, or may not know that it knows how to use it ... The point is that a selforganized system becomes a cognitive system only beyond a minimum number (namely two) of hierarchical levels; (this means that one-to-one mapping does not produce cognition).

The sine qua non prerequisite for the emergence of a selforganizing cognitive system is the ability of this selforganized system to replicate another system (including itself or parts of itself). This self-replication ability is in turn the indication that the system knows how to compress and simulate. Examples: Some chemical reactions constitute dynamical systems which are selforganizing but are not cognitive. The genome of any biological organism on the other hand is a cognitive system (the simplest one) since via transcription and translation (a process involving two hierarchical levels, the level of nucleic acids and the level of polypeptides) it simulates and eventually reproduces itself. Clearly then, cognition is instrumental in compressing the complexity of another system (including parts of itself) via the formation of collective properties (cross correlations or statistical crossmoments) - thereby dramatically reducing the initial number of the degrees of freedom of the system under observation. Alternatively, as we are going to see, an attractor acting on any point of its basin (set of initial conditions) does de facto the same.

Concluding then this introduction, we suggest that communication is involved in triggering the emergence of new spatial-temporal patterns via a sequence of instabilities, producing bifurcations which lead to broken symmetry, and subsequently in creating and maintaining multiple attractors in dynamical systems, which attractors categorize and compress such patterns.

2. What is a Cognitive Channel?

The human mind possesses the unique capability of "mapping" the external (as well as part of the organism's internal) world - that is, it "compresses" long and complex strings of impinging environmental stimuli ("observations") - and then uses these "minimal length algorithms" in order to simulate physical phenomena - thereby revealing the "laws of nature".

We may theorize that this process of "self"-organization and category formation is implemented via a set of coexisting (strange) attractors in the cognizant apparatus - each one of which attracts (and therefore compresses) whole subsets of "initial conditions" the sum-total of which constitute the set of external stimulae (Fig. 2).

This set of initial conditions forms the "basin" of the attractors, and the processes of partition and category formation in the mind involve the topology of the separatrices among the individual subsets of the basin.

Let us then try to sketch the dynamical equivalent of an information channel mediating the mapping between a set of messages of the "transmitter" and a set of "regimes" of the "receiving" - or rather the cognitive apparatus - playing the role of memory bank. (Let us remind ourselves again that an information channel according to Shannon's "static" information theory provides a one-to-one mapping between the transmitter and the receiver, and is in no way associated with the act of "cognition". The sole role of a classical information channel is the mere copying of the transmitted patterns, as they are, at the receiving end with the greatest possible reliability.)

Fig. 2 A cognitive channel

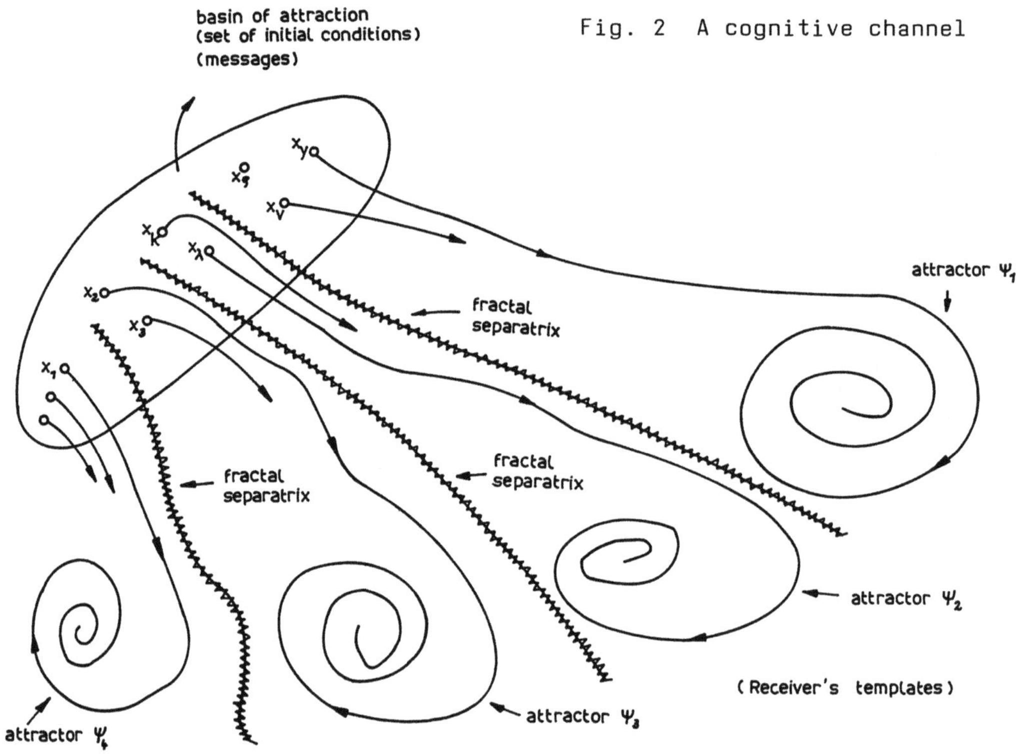

We propose that the set of messages should be identified as a set of initial conditions or the basin of the attractors in an N-dimensional state space where N stands for the number of variables mediating the dynamical phenomenon under consideration.

The receiver, or, more generally, the cognitive apparatus is identified as the set of attractors, of much lower dimensionality, embedded as invariant compact subsets in the same N-dimensional state space.

Falling into the attractor via (short or long) transients in state space corresponds to the act of progressive signal abstraction (or compression). The issue is: what about the dynamical isomorph of the "channel"? or, how is the set of initial conditions, that is the messages, partitioned among different categories, that is among more than one coexisting attractor (steady states, limit cycles, tori or strange attractors)? One has to realize here that the mapping between messages and attractors (or "templates") is not one-to-one: whole subsets (of initial conditions) of the basin converge towards one and only one attractor. Alternatively some points of the basin may undergo a perpetual (or too long) transient trajectory without converging anywhere, or they may even be repelled to infinity. How can we tell which subset of the basin will land into which attractor?

The problem in this general form is, of course, of formidable difficulty. In a very elementary case (two simple attractors, namely two steady states), the issue has been recently addressed. One considers a two-dimensional non-invertible map (the "channel") and finds a fractal separatrix between the two subsets of the basin /12/.

The essential outcome of the above work is that the precision with which a given initial condition is known may play an extraordinary sensitive role on the predictability of the final attractor upon which this initial state will eventually land. This means that an extraordinary high accuracy is sometimes necessary in our measurements for the reliable prediction of the pertinent attractor ("compressor") into which the observed phenomenon will be categorized: clearly then, the set of attractors play in our model the role of dynamically stored patterns in the brain, against which whole sets of externally impinging stimuli have to be "convoluted". (This process may not be accomplished in one single hierarchical step; in a second step the attractors (if numerous at the lower level) will play the role of the members of a hyper-basin towards a new hierarchy of fewer hyper-attractors, and so on.)

Knowing the state space dimensionality of the impinging messages and the (fractal) dimensionalities of the corresponding attractors, one can evaluate in each case the degree of (average) compressibility which a given N-dimensional stimulus is subjected to.

Furthermore, investigating the topology of the separatrices of the basin, one can calculate the uncertainty (or the information gained thereof) associated with the category to which the given stimulus belongs.

Together, the entropy of partitioning and the degree of compressibility give the two essential macroparameters characterizing the cognitive process at a given hierarchical level*.

Once the mapping (that is the "linguistic scheme" adopted by the individual) is known, both the above parameters can in principle be calculated.

We have appreciated already in the introduction the fact that the percentage of compressible sequences in a string of 0's and 1's of length N is extremely small. Considering the 2 a priori equiprobable sequences of length N (the members of the basin) the number of compressible sequences to say K bits is just

$$\Sigma = 2^1 + 2^2 + \ldots + 2^{N-K-1} = 2^{N-K} - 2$$

and the percentage of compressible sequences is

$$\eta = \frac{\Sigma}{2^N} \sim 2^{-K} .$$

So, although in principle there is a numerous set of attractors at each cognitive level, the percentage of those which really deserve the name, that is possess an information dimensionality $c = N - K$ bits ($K \sim O(N)$), is just $\eta = 2^{-K}$, that is insignificantly small.

The moral is that the human mind can simulate a very small subset of natural phenomena. Their overwhelming majority literally "pass through" our heads undetected.

Even so, how come that our templates-attractors really have as basin the set of externally impinging stimuli? To accept this, one has to a priori digest the hypothesis that there exists a "homology" between some patterns of the external world and the categories of the software of the human brain.

But if the human brain is just an unsuspected evolutionary byproduct of the "laws of nature", how come that its software can stimulate the very laws which, acting upon matter, triggered its formation?

* According to our model, the EEG is the pacemaking agent which permutates the attractors on a time division multiplexing basis.

An Example of a Fractal Basin Boundary (Separatrix)

GREBOGI et al. /12/ give the following example of a two-dimensional non-invertible map:

$$\theta_{n+1} = 2\theta_n (\mathrm{mod}\ 2\pi)$$

$$z_{n+1} = \lambda z_n + \cos\theta_n, \quad 2 > \lambda > 1, \quad 0 \leq \theta \leq 2\pi .$$

The Jacobian matrix of this map has eigenvalues 2 and λ. Since both exceed 1, there are no attractors with finite z. In fact, all initial conditions will produce orbits which will asymptotically go either to $z = +\infty$ or $z = -\infty$ as $n \rightarrow \infty$. So we consider $z = +\infty$ and $z = -\infty$ as two coexisting attractors and ask what is their basin boundary. This means: what is $f(\theta)$ such that initial conditions in $z > f(\theta)$ lead the orbits asymptotically to $z = +\infty$ while those in $z < f(\theta)$ lead the orbits to $z = -\infty$?

GREBOGI et al. /12/ starting from the above map, find the expression

$$z = f(\theta) = -\sum_{K=1}^{\infty} \lambda^{-(K+1)} \cos(2^K\theta) .$$

Since $\lambda > 1$ the above sum converges.

Now investigate $df/d\theta$. From the above expression we get

$$\frac{df}{d\theta} = \frac{1}{2} \sum_{K=1}^{\infty} \left(\frac{2}{\lambda}\right)^{K+1} \sin(2^K\theta) .$$

Since $\lambda < 2$ this sum diverges and so $f(\theta)$ is <u>not</u> differentiable. In fact GREBOGI et al. find that the separatrix $z = f(\theta)$ has an infinite length and a fractal dimensionality

$$d = 2 - \frac{\ln\lambda}{\ln 2} = 2 - \log_2\lambda .$$

A final important aspect of the information dimensionality of multiple existing attractors-"classifiers" is the problem of their invariance under smooth coordinate transformations. In other words,we are interested in knowing if the degree of abstraction or compressibility of the basin of initial conditions possesses universality or not. We do not possess a general answer to this question. OTT et al. /13/ have recently investigated the possible invariance of a number of "dimensions" of strange attractors, and have found some specific coordinate transformations ensuring the invariance of only the information dimensionality. Clearly much more work has to be conducted along this direction.

3. Chaotic Dynamics of Pattern Formation

The question "where exactly is information stored?"seems to us to receive a rather original answer from a series of well-known experiments on the perception of random dot interference patterns performed several years ago by GLASS and PEREZ /14/.

The essential conclusions we draw from those experiments is that the information lies neither in the externally impinging pattern (which may be just a random dot aggregation on the plane) nor in the dynamically stored pattern - replica of the receiver (which again may be random); it rather emerges from the flow in the neighbourhood of the <u>singularities</u> - as deduced from the pertinent eigenvalues of the linear or nonlinear, discrete or continuous transformation (mapping or flow) - which transformations the processor <u>imposes between</u> the incoming and the internal pattern.

So one is inclined to consider, at least in a number of cases where such experimental documentation exists, that what we "really" see or cognize in the world is perhaps just the <u>attractors</u> (or the repellers) of the maps or the flows which the processor generates by way of <u>relating</u> external stimuli with internal activity. Biological processors belonging to the same species possess identical nervous systems, and they accordingly tend to generate more or less identical maps or flows - thereby creating as it were by "social consensus" universal pattern classification schemes.

Not only in perception but also in cognition and linguistics, emerging stable patterns may be identified as eigenvalues of non-linear operators, which operators are involved in the recursive character of these systems. For example, recent work in game theory as applied to biology and animal contests /15/ as well as to the process of mutual simulations between competing hierarchical systems each trying to "break" the code of the opponent first, /16,17/ shows that rationality in such tournaments can be completely replaced by evolutionary stability. This points to the emergence of evolutionary stable strategies (E.S.S.) which are point attractors, for the probabilities of moves adopted by the contestants: Such asymptotically stable strategies lead to a stable population polymorphism or stable behavioural repertoires adopted by a single individual.

In linguistics the situation is different since, here, unlike animal or finite state automata games, self-reference is involved. We conclude with two examples, which show that self-consistency, hence logic, is provided also by the existence of stable attractors.

Take as a first example the symbolic self-referential loop:
"This sentence contains $\underline{X_0}$ 0's, $\underline{X_1}$ 1's, $\underline{X_2}$ 2's, $\underline{X_3}$ 3's, $\underline{X_4}$ 4's, $\underline{X_5}$ 5's, $\underline{X_6}$ 6's, $\underline{X_7}$ 7's, $\underline{X_8}$ 8's, $\underline{X_9}$ 9's".
Let the 10-dimensional vector $\underline{X}$ $(X_0, X_1, X_2, \ldots, X_9)$ start from a randomly selected initial condition, and then provide the next iteration by checking the truth of the previous one.

For example start from $X_a(0,1,2,3,4,5,6,7,8,9)$.

The next iteration gives $X_b(2,2,2,2,2,2,2,2,2,2)$.

Next comes $X_c(1,1,11,1,1,1,1,1,1,1)$, followed by

$X_d(1,12,1,1,1,1,1,1,1,1)$ and finally

$X_e(1,11,2,1,1,1,1,1,1,1)$ which obviously

is <u>the</u> attractor corresponding to the selected initial condition and which makes the self-referential sentence above, self-consistent, that is, true. (Through exhaustive search one can find another point attractor as well as the limit cycle of period 2. Here the basin of attraction is partitioned into three subsets.)

As a second example take the sentence:
"This sentence has $\underline{X}$ letters"

Start iterating from X_a = "one"

Then $F(X_a) = 25$ where F is the number of letters after putting in the sentence the given initial condition.

The second "seed" then is X_b = "twenty five" which leads to $F(X_b) = 32$. Likewise X_c = "thirty two" leading to $F(X_c) = 31$ and finally X_d = "thirty one" lands the system to $F(X_d) = 31$ - which is the attractor corresponding to the initially chosen seed "one". There is another point attractor corresponding to a different subset of the basin .

Note finally, that while the truth assessed by the second sentence is "English bound" (translated e.g. into French it becomes false) the first sentence conveys a "translinguistic" truth.

4. Collisions Versus Iterations: A Basic Isomorphism Between Physical and Symbolic Dynamics

The parallelism between a rarefied gas of a finite number of hard spheres and a recurrent algorithm as far as the ergodic and mixing properties are concerned is an excellent example of an isomorphism between hardware and software and provides also a connection between the Kolmogorov-Sinai entropy of the two systems. Consider first the physical system composed of N hard spheres. This system of moving scatterers is sensitive to initial conditions in the following sense:

Take the collision process between a pair of spheres. An angular deviation θ on the trajectory of a sphere changes after a collision to an angle θ' where $\theta' \sim \frac{2\delta}{R}\theta$ where δ is the mean free path and R is diameter of each sphere. Let τ be the intercollisional time (which is of the order of 10^{-10} sec). After a time t the deviation of the trajectory of one sphere in the 6N dimensional state space will grow as $(\frac{2\delta}{R})^{t/\tau}$, so the maximum (positive) Ljapunov exponent of the system is $\lambda \sim \frac{1}{\tau}\log\frac{2\delta}{R}$.

Since the Kolmogorov-Sinai entropy is taken equal to the sum of positive Ljapunov exponents, and since in our system all Ljapunov exponents are positive, the K-S entropy will be $S \sim N\lambda$ where N is of the order of Avogadro's number. Taking into account a typical varified "Lorentz gas" we get:

$S \sim 10^{34} \frac{\text{bits}}{\text{sec}}$ - which represents essentially the new information (or variety) generated by the evolving system per unit time or the information needed to follow the second-to-second trajectory of the system.

Consider next within the unit interval the very simple (piecewise linear) recursive relation $X_{n+1}=2X_n \pmod 1$. Start interacting from any arbitrary initial value X_0 which written in dyadic form is say, $X_0 = .1110101011011..$ Each iteration moves the dot one position to the right. After ξ iterations the value of the number will be

$$X_\xi = 2^\xi X_0 .$$

The initial value X_0 can never be known exactly - even in theory. Why? Because this would amount to asking an infinite number of questions (and getting an infinite number of answers in the yes-no form), that is, it would amount to performing an infinite number of measurements. The entropy of a continuous distribution is infinite. Given that each measurement requires a finite time (commensurate with the switched time or the relaxation time of the particle-composed "decision making" element of the receiver) getting knowledge of any number X_0 with infinite accuracy would require an infinite amount of time.

Now $(\Delta X_\xi) = 2^\xi (\Delta X_0)$ which means that the initial uncertainty ΔX_0 (however small, yet finite) is amplified exponentially with the number of iterations. If X_0 is known with an accuracy $\sim 10^{-\tau}$ after $\xi_c = \tau \log_2 10$ iterations the initial condition X_0 is completely forgotten, and from then on the recurrent algorithm functions as a "pseudo-random generator" producing an essentially incompressible sequence of digits - although the relationship responsible for such a sequence is quite deterministic.

The K-S entropy here equals $\log_2 2=1$ bit/iteration. This "software" system is also ergodic and mixing.[2]

References

1. J.S. Nicolis: Kybernetes 11, 123 (1982); 11, 269 (1982); 14, 167 (1985)
   D. Gikas, J.S. Nicolis: Z.Phys.B 47, 279 (1982)
2. J.S. Nicolis, J. Milias-Argitis, C. Carabalis: Kybernetes 12, 9 (1983)
3. J.S. Nicolis: J.Franklin-Inst. 317, 289 (1984)
4. J.S. Nicolis, G. Mayer-Kress, G. Haubs: Z.Naturforschung 38a, 1157 (1984)
5. J.S. Nicolis, I. Tsuda: Bull.Math.Biol. 47, 343 (1985)
6. J.S. Nicolis: Dynamics of Hierarchical Systems, Springer-Series in Synergetics, Vol. 25 (Springer, Berlin Heidelberg New York 1985)
7. J.S. Nicolis: Chaotic Dynamics Applied to Information Processing, Rep.Progr.Phys., to appear; Dissipative Structures, Rep.Progr.Phys., to appear
8. P.W. Anderson: Science 177, 393 (1972)
9. G. Nicolis, I. Prigogine: Self-Organization in Non-Equilibrium Systems (Wiley, New York 1977)
   G. Nicolis, I. Prigogine: Proc.Nat.Acad.Sci. USA 78, 659 (1981)
10. H. Haken: Synergetics, Springer-Series in Synergetics, Vol. 1 (Springer, Berlin Heidelberg New York 1978)
11. G. Chaitin: Sci.Amer. 47 (1975)
12. C. Grebogi, E. Ott, J. Yorke: Phys.Rev.Lett. 50, 935 (1983)
    C. Grebogi, S. McDonald, E. Ott, J. Yorke: Phys.Lett. 99A, 415 (1983)
13. E. Ott, W.D. Witters, J.A. Yorke: J.Stat.Phys. 36, 687 (1984)
14. L. Glass, R. Perez: Nature 246, 360 (1974)
15. John M. Smith: Evolution and the Theory of Games (Cambridge Univ. Press, Cambridge 1982)
16. J.S. Nicolis, E.N. Protonotarios, M. Theologon: Int.J.Man-Machine-Studies 10, 343 (1978)
17. J.S. Nicolis, E.N. Protonotarios: Int.J.Bio-Medical Computing 10, 417 (1979)

The Evolutionary Triad

M.V. Volkenstein

Academy of Sciences of USSR, Institute of Molecular Biology, SU-117 312 Moscow, USSR

1. Introduction

Darwin's theory of evolution belongs to the greatest achievements of scientific thought. For the first time in the history of natural sciences the mechanism of the formation of an ordered, directed process - the biological evolution - in a system with randomly varying properties was established.

By this, Darwin developed the principles of synergetics in the case of living nature a long time before its general foundation. When Boltzmann was asked how he would call the 19th century, he answered: "I would call it the century of Darwin." Darwin's theory is not a completed, rigid doctrine, of course the theory of evolution is evolving itself and undergoes changes. In the first half of this century the fusion of Darwinism and genetics took place. The synthetic theory of evolution was developed which represents a more fundamental proof of Darwinism and helped to overcome problems Darwin was confronted with, as, e.g., The "nightmare of Jenkins". The history of this fusion is magnificently described /1,2/. Currently we experience a new stage in the development of science. New branches of science with direct relation to evolution, like molecular biology, theory of information and, in particular, synergetics, have emerged. For the theory of evolution, the time is at hand to incorporate these branches. The first step in this process is well under way; the present article is dedicated to it. We start with synergetics. Today we understand completely that any process of selforganization of an evolving system, cosmos as well as biosphere, takes place by export of entropy into the surrounding medium /3/. Evolutionary systems are open and far from equilibrium. Consequently, they develop unstable states which are left by the system in favour of comparatively more stable states as a result of bifurcations. In what follows, new instabilities and new states arise, etc. Of course, these processes are connected with changes of the state of the surroundings to which the evolutionary system itself belongs. We are concerned here with networks of mutually connected Markov processes, the stochastic matrices of which change with time.

This is a very general concept. Important reasons suggest that the formation of a new structure, i.e. the creation of new information as a result of the remembrance of a random choice /4/, has always the character of a nonequilibrium phase transition. Hence these processes have to be considered on the basis of synergetics /5/, which with the same right could be defined as the theory of dissipative systems /6/. In open systems far from equilibrium, small fluctuations can grow up to macroscopic scales. Contrary to the claims of MONOD /7/, the structure of living systems is in principle different from the equilibrium structure of a rock, a dead crystal. The aperiodic crystal of SCHRODINGER /8/ is not only different from a periodic crystal because of its very much smaller content of excess information, but, first of all, thanks to the remoteness of a living system from equilibrium.

We have dealt with the information aspect of the theory of evolution in other publications /9-12/.

2. Punctualism

It became apparent recently that one thesis of classical Darwinism has to be reconsidered. This is not surprising, since the theory of evolution is itself an evolving system, changing with time. The problem under consideration is the phyletic gradualism - the permanent accumulation of small changes in the populations which give rise to species formation and macroevolution.

As has been demonstrated convincingly in the works of GOULD and ELDREDGE (cf. /13/) the formation of species occurs via the development of a punctuated equilibrium. In other words, natura facit saltus. The main argument states, that according to the palaeontological chronicle new species, and even more the higher axons, emerge in a comparatively short time after a long period of stagnation. The stagnation can take millions of years, whereas the formation of a species takes place within some ten thousand years. This has been clearly demonstrated by WILLIAMSON /14,15/ by investigating the palaeontology of several invertebrates of the sea. The concepts of punctualism form the basis of the modern theory of macroevolution /16/.

MAYR disqualifies the punctualism, because it would correspond to the idea of GOLDSCHMIDT of a hopeful monster. The basis for this consists in the claim by GOULD that "macroevolution proceeds by the rare successes of these hopeful monsters" /17/. As we will see shortly, the virtual intention of punctualism is different. In the paper /18/ (see also /9,19,20/) it is shown that a simple biallelec system diverges, similar to a phase transition of first or second order. In other words, the formation of new species is the result of a non-equilibrium phase transition of a synergetic event. The intermediate stage proves to be unstable.

This is a thesis of phenomenological thermodynamics; neither the scale or the kinetics of the phase transition will be considered here. Consequently several cases of punctuated equilibrium can hardly be distinguished from phyletic gradualism - the phase transition may be smeared or dilated in time. Nevertheless it always takes a time very much shorter than the duration of the stagnation.

Which kind of genes are responsible for the formation of species and macroevolution? The genes which control the work of the structural genes. Today it is possible to assume that the regulation of genes results from the function of so-called homeotypic genes and from a shift of the genes inside the genome. The mutations of homeotypic genes give rise to more significant alterations of the structure of the organism. The mutation in Drosophila called Antennapedia, e.g., results in the formation of an additional pair of legs instead of antenna. This in indeed a hopeful monster (cf. /1/). The mobility of the genes, which has been thoroughly investigated in recent years, gives rise to the formation of new genetic blocks and, apparently, makes macroevolution possible.

3. "What for?" and "Why?"

In biology frequently the question is posed: "What for?". By this is meant that this or that characteristic, this or that process have an adaptive meaning. Why did the Tyrannosaurus have such small forelegs and such big hind extremities? - What were these small hands necessary for; Obviously to tickle the female, answers GOULD ironically.

The question "What for?" has virtually no scientific meaning. Similarly, we do not state in physics that the ions Na^+ and Cl^- combine with the intention to form a cubic crystal. Scientific sense has only the question "Why?", in physics as well as in biology.

The literature (cf. /21/) discusses some peculiarities of the hyena Crocuta crocuta. The outward genitals of the female of this species do not differ from those of the male. The question "What for?" arises. The question is senseless, there is no adaptive meaning of this peculiarity. Nevertheless the question "Why?" can be answered. In this hyena the female is significantly taller than the male, which is caused by an increased content of androgenic steorid hormones.

What did the Cro-Magnon, the brain of which did not differ from that of the modern human, need the ability of abstract thinking for, which after all made the construction of the A-bomb possible? It had no adaptive meaning. One could think that the evolution of the brain of Homo sapiens is a byproduct.

What can be said finally about the adaptive meaning of the existence of four extremities, a characteristic common for all terrestrical vertebrates? The only exceptions, angels and centaurs, are not considered scientifically. Is it possible to assume that four extremities are more convenient than six or eight?

In many cases no answer for the question "What for?" exists. The question "Why?" will be answered sooner or later. The existence of four extremities of terrestrial vertebrates, e.g., results from the fact that their predecessors, the Latimeria, had four corresponding limbs. Evolution is irreversible and directed. It is not only directed because of natural selection, but also because of its inherent limits. Such inherent limits are the already formed structure of the organism, and the possible methods of its variations. Whether one interprets this simply as a limitation of the outer selection, or whether one should speak of independent inner factors of the irreversible evolution, is a purely semantic question. Anyway, the creative possibilities of natural selection are always severely limited. This was splendidly realized by DARWIN when he wrote /22/: "... we clearly see that the nature of the conditions is of subordinate importance in comparison with the nature of the organism in determining each particular form of variation - perhaps of not more importance than the nature of the spark - by which a mass of combustible matter is ignited, has in determining the nature of flame."

From what has been said follows that a number of characteristics have no adaptive meaning, are not the result of natural selection. Situations are possible where the considered characteristic had an adaptive meaning in previous stages of evolution and lost it later. And, vice versa, a nonadaptive characteristic can become an adaptive one in the following time.

The non-adaptivity, e.g. the back of an adaptive meaning of a number of characteristics, has been favoured in the papers /23,24/. We realize that non-adaptivity follows from the limitations of natural selection. At the same time it follows immediately from punctualism. The formation of new species as the result of a phase transition implies the occurrence of population with several not yet adapted characteristics. The complete adaptation cannot be implemented in the short period of species formation.

The theoretical and experimental consideration of phase transitions in biological evolution has just begun. The investigation of the thermodynamics and kinetics of nonequilibrium phase transitions in homogeneous and distributed systems will be carried out in near future. These investigations should be based on the results of modern genetics and molecular biology.

4. The Molecular Basis of Evolution

On the molecular level of living systems, the nucleic acids - DNA and RNA - carry out the legislative power whereas the proteins form the executive power, mostly via their fermental function. Incidentally, it has been shown recently that also some kinds of RNA take part in fermental catalysis. Today the primary structure, i.e. the sequence along the chains, is known for a large number of proteins and nucleic acids. The comparison of primary structures of homologous proteins of different organisms shows that the "texts" of these proteins are the more similar the nearer the organisms are located to each other in the evolutionary tree. The cytochrome C of a human, e.g., differs from that of a dog by 11 amino acids and from that of yeast by 45 amino acids. In this way the comparison of the primary structure of biopolymers makes it possible to find exact measures for the distance betwen different species. These measures are relative, taken for different proteins they do not coincide completely. Notwithstanding, proteins of more-different species differ more than those of related ones. It is tempting to assume that these differences were caused by natural selection on the molecular level; in other words, any alternation of an amino acid in a protein caused by mutation (i.e. corresponding to an alternation of one or several nucleotides in the DNA) has adaptive meaning. The molecular basis of the formation of species consists in the changes of genes and the corresponding proteins.

However, as already stated, not all characteristics are adaptive. The claim that the differences between the species are due to the differences of their proteins has only limited sense. In the paper /25/ it has been shown that 44 functional proteins of human and chimpanzee differ with respect to their content of amino acids by only 1%. The differences between the species are not predominately determined by the structure of the proteins but by the regulation of their synthesis. Of importance are the quantity of the protein produced, time and location of its occurrence in the evolving organism, i.e. the exact answers to the questions how much? when? and where? The formation of species and macroevolution are determined on the molecular level by changes in the regulation of genes and by the formation of protein, not present in the preceding forms.

The above also follows from punctualism, from the existence of nonequilibrium phase transitions in the evolution. Alterations of proteins by mutations are not phase transitions but gradual changes, replacing one amino acid after another, no abrupt reconstruction of the whole primary structure takes place. The biological function of a protein is not uniquely determined by its primary structure but by the spatial construction of the macromolecule. If there were no correlation between the primary and the spatial structure, molecular biology and molecular genetics would lose their sense.

There are of course correlations, however, they are not unique, they are degenerate.

In the paper /27/ it has been shown that 9 globins, among them the hemoglobin and the myoglobin, which differ greatly from each other in their primary structure, have a very similar spatial structure. The situation is similar for cytochrome C and other proteins. It is well known that averages over the primary structure of globin proteins give no significant deviations from a random distribution of amino acids. PTITSYN has shown that this remains true also for higher levels of organization of proteins.

In a protein globule the chain forms elements of a secondary structure: α- helices, β- ligaments, and disordered parts. PTITSYN has shown that the length distribution of such elements coincides with that of a statistical co-polymer with random distribution of links. This is true as well for the cluster formed by the α- and β-

elements. The protein represents an "edited statistical co-polymer", the structure of which is the result of the remembrance of a random selection (cf. /29/). The protein macromolecule is a complicated machine and consists, roughly speaking, of an active, working part (say the active centre of the ferment) and of a "passive part", a moveable carcase enclosing the active part. The "editing" is mostly done in the active part by natural selection. PERUTZ has shown that the evolutionary adaptation in hemoglobins consists in only very few (one to five) alterations of amino acids in the active centre /30/. Consequently an important part of alterations caused by mutations which are observed in proteins are not adaptive. This is in accordance with the above concepts and with punctualism.

5. The Neutral Theory of Molecular Evolution

Independent of the quoted considerations, KIMURA has developed the so-called neutral theory of molecular evolution /31,32/. According to this theory, a substantial part of alteration in the proteins and nucleic acids (but not all!) are neutral, they do not influence the biological function of the polymers and are not subject to the selection pressure.

The arguments of this theory are based on the observation of the evolution of proteins.

KIMURA has shown that the frequency of replacements of amino acids for a given protein is approximately the same for different lines of evolution. It does not depend on the nature of the species or the magnitude of the population. Different proteins evolve with different velocities. Given by the number of alterations per amino acid in the course of 10^9 years this evolutionary velocity is, e.g., 9.0 for fibrinopeptides, 1.4 for hemoglobin, 0.3 for cytochrome C and 0.006 for histone 4, where the evolutionary velocity of the active part is considerably less than that of the passive part.

With the help of the method of gel-electrophoresis, the polymorphism of many proteins was established. In one and the same organism a particular protein can occur in many forms, different in their primary structure but very similar in their function. The polymorphism of proteins also indicated the neutrality of mutations.

Physically, the neutral theory results in the degeneracy of the correlation between primary and spatial structure of the protein already mentioned. We have already given arguments in favour of neutralism different from those of KIMURA. They should be completed by the following: As shown by OHTA, mutations leading to a small drawback behave similarly to neutral mutations /32/. The criterion for the smallness is a small value of the selection coefficient.

The genetic code shows a special stability against perturbations (/23/, see also /10,11,34/). In the case of a single change of nucleotides in the DNA, predominantly amino acids with similar properties are replaced. That is an additional argument in favour of neutralism.

Unlike machines constructed by man, the living organism is far more labile. It possesses sufficiently broad tolerances and backlashes. That is determined by the nature of the organism as a chemical machine. In artificial machines, made of solid materials, a worn and torn part or one not of appropriate quality has to be replaced. In an organism or in a cell, different methods of compensation of inadequate parts are possible. If, e.g., as a result of a mutation some ferment catalyses a reaction with lower velocity than necessary, a correction of the mutation by a feedback mechanism will be possible on the level of metabolism as well as on the level of genetics. The simplest example is the increasing production of a deteriorating ferment /35/. The compensations support the homeostasis. Compensated mutations behave obviously like neutral ones.

Finally, it should be emphasized that, if any amino acid of a protein had to be fixed by natural selection, the lifetime of the universe would have been totally insufficient for the formation of the existing proteins and the existing biosphere.

6. Punctualism, Non-Adaptivity, Neutralism

We see that the three peculiarities of biological evolution form a triad of mutually connected phenomena

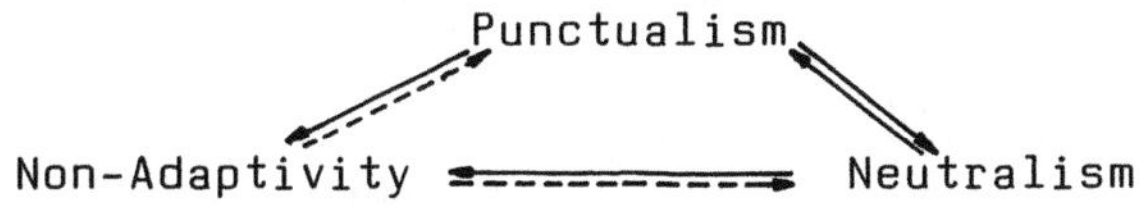

Non-adaptivity and molecular neutralism follow logically from punctualism. Neutralism means absence of adaptivity. The corresponding arrows in the figure are indicated by heavy lines. Non-adaptivity of several characteristics can result in neutralism and punctualism, but need not necessarily do so. These arrows are shown with broken lines.

All the features arise from the synergetic nature of organisms, from the inevitable occurrence of phase transitions in dissipative systems. The beginning of an ordered evolution out of chaos on the basis of natural selection, the Darwinian synergetics, results inevitably in punctualism and consequently in the related non-adaptivity of several characteristics and in the neutrality of several mutations on the molecular level. The analogy between the formations of species and phase transitions has been stressed first by BERG, who wrote in 1921 /36/:

"Birth and death of individuals, species, ideas are catastrophic processes. Before these categories begin to exist there is a long invisible period, governed by particular laws, and after it an immediate leap, saltus, occurs, which establishes their existence, their spreading and their fight for a 'place under the sun'. The transition of a gas in a liquid is a discontinuous change."

The orthodox darwinists are frequently willing to accept punctualism and the non-adaptivity of a number of characteristics but rise against the neutral theory. But Darwin spoke about organisms and populations, not about molecules. There is no contradiction between the discussed triad and Darwinism. There are reasons to believe that punctualism, non-adaptivity and neutralism are further developments of the theory of evolution. These concepts make a new synthesis visible - the inclusion of the ideas of synergetics and molecular biology into the theory of evolution. The fundamental concepts discovered by Darwin, however, remain unchanged.

References

1. E. Mayr, W. Provine: The Evolutionary Synthesis (Harvard Univ. Press, Cambridge 1981)
2. E. Mayr: The Growth of Biological Thought: Diversity, Evolution and Inheritance (The Belkaap Press of Harvard Univ., Cambridge 1982)
3. W. Ebeling, R. Feistel: Physik der Selbstorganisation und Evolution (Akademie-Verlag, Berlin 1982)
4. H. Quastler: The Emergence of Biological Organization (Yale Univ. Press, New Haven 1964)
5. H. Haken: Synergetics (Springer, Berlin Heidelberg New York Tokyo 1983)
6. P. Glansdorff, I. Prigogine: Thermodynamic Theory of Structure, Stability and Fluctuations (Wiley, London New York Sydney Toronto 1971)

7. J. Monod: Le Hasard et la Necessité (Senit, Paris 1970)
8. E. Schrödinger: What is Life? (Cambridge 1944)
9. M.V. Volkenstein: General Biophysics, vol. 2 (Academic Press, New York, London 1983)
10. M.V. Volkenstein: Biophysics (Mir, Moscow 1983)
11. M.V. Volkenstein: Physics and Biology (Academic Press, New York London)
12. M.V. Volkenstein, D. Chernavsky: in: Self-Organization. Autowaves and Structures Far from Equilibrium (Ed.: V. Krinsky) (Springer, Berlin HeidelbergNew York 1984)
13. S. Gould, N. Eldredge: Paleobiology 3, 115 (1977)
14. P. Williamson: Nature 294, 214 (1981)
15. P. Williamson: Nature 293, 437 (1981)
16. S. Stanley: Macroevolution. Pattern and Process (W. Freeman and Co., San Francisco 1979)
17. S. Gould: Nat.Hist. 86, 22 (1977)
18. B. Belintsev, M.V. Volkenstein: Doklady Acad.Nauk USSR 235, 205 (1977)
19. M.V. Volkenstein: J.Theor.Biol. (in press)
20. M.V. Volkenstein: Uspekhi Physicheskikh Nauk 143, 429 (1984)
21. J. Bonner (Ed.): Evolution and Development (Springer, Berlin Heidelberg New York 1982)
22. Ch. Darwin: Origin of Species. Chapter 1
23. S. Gould: Science 216, 380 (1982)
24. S. Gould, E. Vrby: Paleobiology 8, 4 (1982)
25. M. King, A. Wilson: Science 188, 107 (1975)
26. M.V. Volkenstein: J.Obshishei Biologii 42, 690 (1981)
27. A. Lesk, C. Chothia: J.Mol.Biol. 136, 225 (1980)
28. O. Ptitsyn: in: Conformations in Biology (Eds.: R. Srinivasan, R. Sarma) (Academic Press, New York 1983)
29. O. Ptitsyn, M.V. Volkenstein: J.Biomol.Structure and Dynamics (in press)
30. M. Perutz: Mol.Biol.Evol. 1, 1 (1983)
31. M. Kimura: Sci.American 241, Nr. 5, 95 (1979)
32. M. Kimura: The Neutral Theory of Molecular Evolution (Cambridge Univ.Press, London New York Melbourne Sydney 1983)
33. M.V. Volkenstein: Biochim. Biophys.Acta 119, 421 (1966)
34. M.V. Volkenstein: Molecular Biophysics (Academic Press, New York London 1977)
35. M.V. Volkenstein: Molekularnaya Biologia 18, 55 (1985)
36. L. Berg: Nomogenesis or Evolution Determined by Law (MIT Press, Cambridge 1969)

Part V

Chaotic Processes

Fully Developed Chaos in One-Dimensional Discrete Processes

G. Györgyi[1] *and P. Szepfalusy* [2;1]

[1]Institute for Theoretical Physics, Eötvös University, P.O. Box 327, H-1445 Budapest, Hungary

[2]Central Research Institute for Physics, P.O. Box 49, H-1525 Budapest, Hungary

Abstract

It has turned out in recent years that dissipative dynamical systems exhibiting chaotic motion can often be modelled by one-dimensional maps, $x_{n+1} = f(x_n)$, where f(x) is typically a single hump function. An important feature of such maps in the chaotic region is the sensitivity to the change of the usual bifurcation parameter. One can introduce, however, parameters specifying the map, whose variation (at least within a certain range) does not destroy the chaotic state and leads to smooth parameter dependence of the statistical properties. As an example for such family of maps fully developed chaotic 1-D maps are considered, which generate ergodic processes on an interval mapped everywhere two-to-one onto itself.

To carry out the calculations, a perturbation theoretic method is worked out. After general considerations two examples are treated in detail, namely $f_I(\varepsilon,x) = 1-(1-\varepsilon)(1-2x)^2 - \varepsilon(1-2x)^4$ and $f_{II}(\varepsilon,x) = 1-(1-\varepsilon)|1-2x| - \varepsilon(1-2x)^2$. Here ε measures the deviations from the unperturbed maps which are the logistic and the tent maps, respectively.

In this lecture we focus our attention on the ε-dependences of characteristic inverse time constants, namely the average Ljapunov exponent (λ), which agrees with the Kolmogorov-Sinai entropy for such maps, the relaxation rate (r) of probability distributions on the attractor and a third one (γ), which arises by approaching the Kolmogorov-Sinai entropy through a series of truncated entropies. Larger characteristic rates in general are regarded to indicate larger erraticity in the system. While λ and γ assume their maximal values at $\varepsilon = 0$ ($2\lambda_0 = \gamma_0 = \ln 4$) with a first correction of $O(\varepsilon^2)$, r is found to exhibit a linear ε-dependence for small ε ($r_I = \ln 4 + \varepsilon - \varepsilon^2/2 + \ldots$, $r_{II} = \ln 4 + 3\varepsilon - \varepsilon^2/2 + \ldots$) and reaches its maximal value at parameter settings not distinguished by the other characteristic rates. It is shown that these differences in behaviour of the characteristic rates are generally valid for fully developed chaotic maps. Furthermore, the behaviour is studied at intermittent fully development chaotic maps.

References

G. Györgyi, P. Szepfalusy: J.Stat.Phys. 34, 451 (1984)
G. Györgyi, P. Szepfalusy: Z.Phys.B-Condensed Matter 55, 179 (1984)
G. Györgyi, P. Szepfalusy: Phys.Rev. A31, 3477 (1985)
P. Szepfalusy, G. Györgyi: to be published

Auto-Oscillatory Regimes in the Chain of Coupled Generators

V.S. Anishchenko

Saratov State University, N.G. Chernyshevsky, SU-Saratov, USSR

Investigations of oscillations in the chains of coupled generators are of great interest from the points of view of both modelling active media and different uses /1/. Judging by the available data, the effects observed in similar systems can be grouped about two opposite phenomena: complete mutual locking of the elements of the "medium" and spatially homogeneous chaos. However, we think that the partial locking regime, complicated oscillation-locking regime, as well as regimes of spatially non-homogeneous chaos must naturally exist in continuous and discrete active media. It is very important to reveal and analyze such regimes for understanding in nature of forming coherent dissipative structures in active and excited media and transition from regular structures to chaos.

In this paper we give the results of the experimental and numerical investigations of spatial locking processes and chaotization of oscillators as an example of the chain of one-directedly coupled generators with inertial nonlinearity /2/. For the first time the following effects were observed: 1 - locking of the generators of the chain in subharmonic oscillation regimes, 2 - spatial (along the chain) transition in chaos through succession of period-doubling bifurcations, 3 - spatial bifurcations of chaos development resulting in the increase and subsequent saturation of intensity of broadband chaotic pulsations (chaotic locking).

The chain investigated included 10 identical generators with one-directed resistive coupling described by the following equation:

$$\dot{x}_j = mx_j + y_j - x_j y_j + \alpha x_{j-1}, \quad \dot{y} = - x_j \qquad (1)$$
$$j = 1,2,\dots,10, \quad x_{j-1} = 0 \text{ for } j = 1$$

$$\dot{z}_j = g\,[I(x_j)x_j^2 - z_j], \quad I(x_j) = \begin{cases} 1, & x_j > 0 \\ 0, & x_j \leqslant 0 \end{cases},$$

where m is the parameter of excitement of partial generator, g is the parameter of inertiality, α is the coefficient of coupling. In physical experiment the parameter m was changing within the limits of $0 \leqslant m \leqslant 1.5$ simultaneously for the whole chain, and the parameter g was fixed at a level of 0.3. Measurements were performed for different values of the parameter of coupling: $0.05 \leqslant \alpha \leqslant 0.3$. An important feature of the chain was that with varied parameter m, autonomous partial generator is able to realize different types of regular and strange attractors, and thus modelling the extent of nonequilibrium of elementary cells of nonlinear "medium" /2,3/.

The diagram of regimes in the plane of parameters $k \sim m$ for the chain (they were continuously changing) and j (a discrete analogue of the "spatial" coordinate), represented in Fig. 1, whows main bifurcation phenomena in the system.

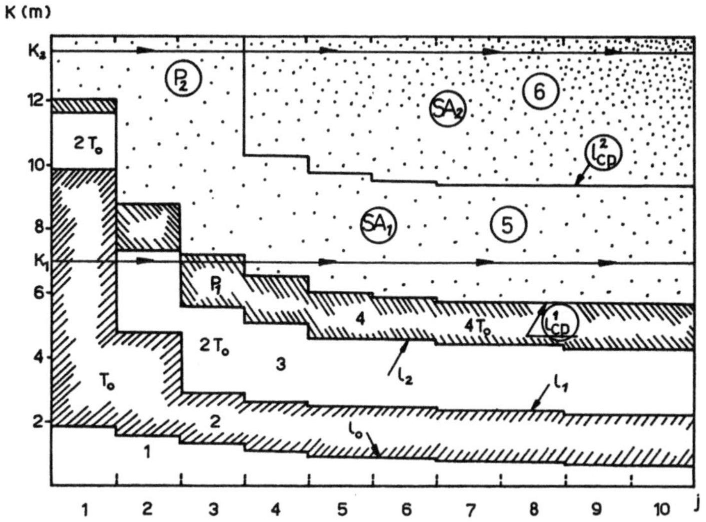

Fig. 1 Experimental bifurcation diagram of oscillation regimes in system (1) for $\alpha = 0.1$ and $g = 0.30$

This diagram shows six regions with different regimes limited by corresponding bifurcation lines l_k. In the region 1 the system is not excited. The line l_0 characterizes smooth bifurcation of ANDRONOV-HOPFF and in the region 2 the system generates periodical one-path oscillations (period of cycles $\simeq T_0$). As for the regions 3 and 4, the regimes of subharmonic oscillations are also realized there, when the lines l_1 and l_2 are crossed, the period-doubling bifurcations $T_0 \rightarrow 2T_0$ and $2T_0 \rightarrow 4T_0$ are observed correspondingly. Transition across the line of critical values of parameters l_{kp}^1 is accompanied by the smooth appearance of a banded strange attractor SA_1, which is evolved then into spiral-type attractor SA_2. This process corresponds to the smooth transition from the region 5 to the region 6 across bifurcation line l_{kp}^2. When crossing l_{kp}^2, in power spectra of oscillations δ -shaped spikes on the basic frequency harmonics $f_0 = T_0^{-1}$disappear, while in the region 5 near l_{kp}^2 they can still be seen.

With motion along the chain with different fixed k, the following situations are possible. For $1 < k < 2$ along the chain $j \geqslant 2$ locked regime of one-path periodic oscillations was established. In the interval $3 < k < 4.5$ the generators $j \geqslant 3$ are locked in the 2-cycle regime and in the 4-cycle regime for $4.5 < k < 5.5$ and $j \geqslant 5$. 8-cycle regimes in the experiments were registered in the narrow regions and not in all generators of the chain. With increase of $k \simeq 10$ in generators $j \geqslant 2$ the banded attractor SA_1 was realized, but for $k > 12$ in most generators of the chain developed stochasticity SA_2 was observed. Fi-

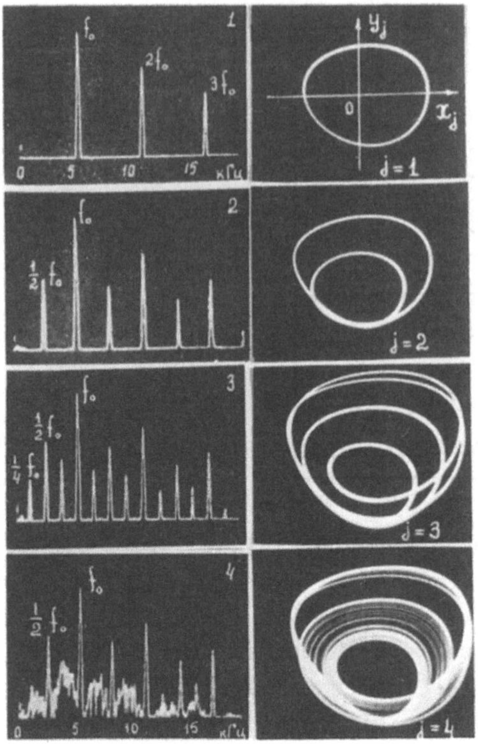

Fig. 2 Transition to chaos through succession of spatial period-doubling bifurcations. Physical experiment for α=0.1, g=0.3 and k_1=6.9

gure 2 represents the evolution of regimes along the chain for the value of $k_1 \simeq 7$ (motion along the line P_1 in Fig. 1) in the form of oscillograms of power spectra S_f of realizations $x_j(t)$ and corresponding projections of phase trajectories in the plane of variables $x_j y_j$. In the first three generators of the chain succession of "spatial" period-doubling bifurcations $T_0 \rightarrow 2T_0 \rightarrow 4T_0$ is determined. In the fourth generator there appears the two-path band of Feigenbaum's attractor SA_1 which evolves along the chain $j \geq 5$ into a one-path band whose power spectrum is continuous with δ-shaped spikes at frequencies nf_0, n = 1,2 and 3. The number of period-doublings along the chain is always finite, which is conditioned for the discreteness of the system along the spatial coordinate j. In the distributed system the complete cascade of period-doubling bifurcations limited only by the fluctuation level is quite possible.

Let us consider the evolution of regimes along the chain for k_2 = 13.5 (see the direction of P_2 in Fig. 1). Without coupling, partial generators of the chain are in a regime of developed one-path band of the attractor SA_1. Figure 3a indicates spectra of oscillation

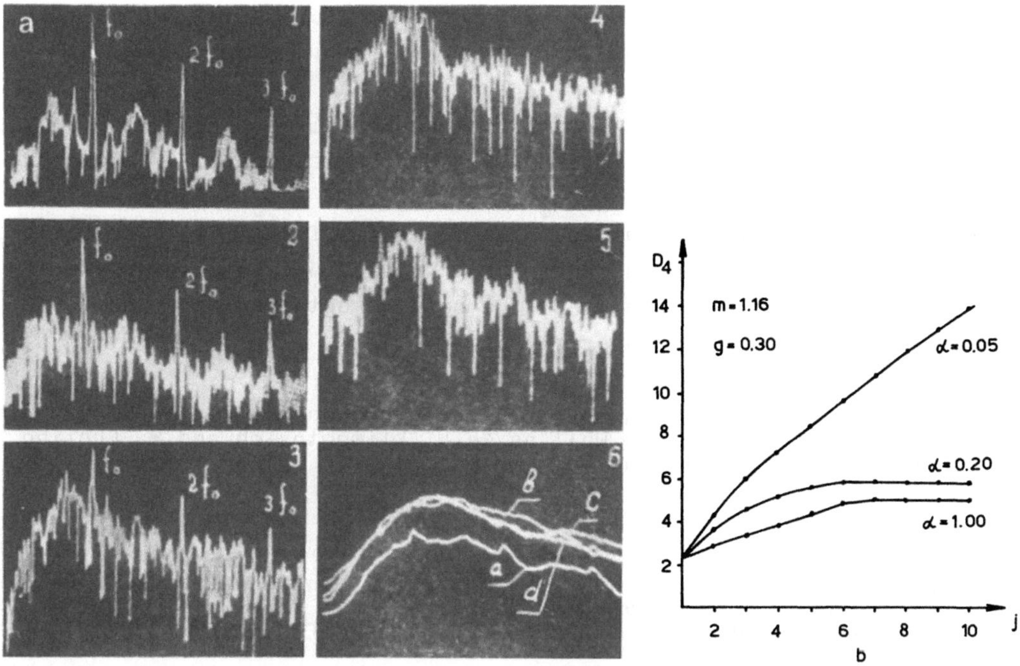

Fig. 3 Effect of chaotic locking, a - evolution of power spectra along the chain for K_2 = 13.5, on photo 6 average spectra in generators j = 1, 4, 6 and 10 (correspondingly to a, b, c, d); b - Ljapunov dimension of attractors for different coefficients of coupling

power in the first five generators of the chain (see photos 1-5). In generators j=2,3 the topology of SA_1 is unchanged, but for j=4 δ-spikes at frequencies nf_0 disappear; it means that the transition into regime of SA_2 has been realized. In generators $j \geq 5$ power spectrum is practically unchanged, which suggests the effect of saturation in chaos; this effect can naturally be called the chaotic locking /4/.

The effect revealed is the result of nonlinear processes of fractal dimension stabilization of SA_2 along the chain that is confirmed by numerical experiments. Figure 3b represents the results of calculations of Ljapunov dimension D_4 by KAPLAN-YORKE performed with the help of special algorithm elaborated for similar classes of problems, which reduces the calculation time of the whole spectrum of Ljapunov exponents. As is evident from the graphs with weak coupling α=0.05 dimension D_4 along the chain increases nearly in a linear law /5/. Structure of stochasticity also becomes more complicated, which is confirmed by the evolution of all basic characteristics of the attractors. However, with coupling degree the dimension is gradually stabilized along the chain and reaches the constant and lower level more quickly.

The graph for $\alpha = 0.2$ shows that for generators with $j \geq 6$ D_4 is equal to 5.85 which is much less than the dimension of phase space of the system.

Investigations performed confirm that for fixed value of parameters $\alpha = \alpha_0$, g = const in the system (1) there exist critical values of nonequilibrium parameter $m = m_0$, and for all $m < m_0$ effects of locking in one of the regular regimes takes place; as for $m > m_0$ bifurcations resulting in spreading and subsequent saturation along the chain of stochastic "wave" are realized. The results obtained were theoretically explained in the paper /4/.

The author thanks M. Saphonova, T. Letchford and D. Postnov for their help in experimenting, as well as Yu.L. Klimontovich and W. Ebeling for their attention to the work.

References

1. A.V. Gaponov-Grekhov, M.I. Rabinovich, I.M. Starobinets: JETP Lett. 39, 561 (1984)
2. V.S. Anishchenko, V.V. Astakhov: Radiotekh.Electron. 28, 1109 (1983)
3. V.S. Anishchenko, V.V. Astakhov, T.E. Letchford, M.A. Saphonova: Izvestiya Vuz.Radiofizika 26, 832 (1983)
4. V.S. Anishchenko, I.S. Aranson, D.E. Postnov, M.I. Rabinovich: Dokl.Akad.Nauk SSSR (1985) (in print)
5. V.S. Afraimovich, M.I. Rabinovich, V.I. Sbitnev: JETP Lett. 11, 338 (1985)

Universal Scaling Properties of Nonlinear Dissipative Systems

R. W. Leven and B. Pompe

Ernst-Moritz-Arndt-Universität Greifswald, Sektion Physik/Elektronik, Domstraße 10a, DDR-2200 Greifswald, GDR

1. Universality in One-Dimensional Mappings

As was pointed out by different authors, there is a striking resemblance of chaotic phenomena to critical phenomena (see e.g. /1/). In particular the concepts of scaling and universality are crucial for the understanding of both. The most interesting universal properties of chaotic systems are connected with transitions from ordered to chaotic motions and vice versa, as well as transitions from one chaotic regime to another. Such transitions occur if a control parameter is varied, and in many cases the system behaviour is complicated near the critical point where the transition occurs.

The most investigated and best known transition phenomenon in chaotic motions is perhaps the period-doubling route to chaos in one-parameter systems. For one-dimensional maps

$$x_{n+1} = f(x_n, r) \tag{1}$$

this cascading behaviour in the periodic regime has been shown to display universal features characterized by a functional renormalization group equation and two universal ratios, the rate of bifurcation, $\delta = 4.6692...$, and the scale reduction factor

$\alpha = 2.5029...$ (see e.g. /2/).

Above the value r_∞ of the parameter where the sequence of period doubling bifurcations accumulates, one finds a reverse sequence of bifurcations at parameters $\tilde{r}_k$ where 2^k bands of a chaotic attractor merge into 2^{k-1} bands. The sequence of reverse bifurcations is characterized by the same self-similarity rule as the bifurcation sequence, and obeys the same universal equations with the same universal convergence and scaling numbers δ and α .

2. Ljapunov Exponents and Transition Phenomena

An important notion in nonlinear dynamics is that of sensitive dependence on initial conditions. In the chaotic regime, two trajectories in phase space that initially differ by a small amount will separate exponentially in time, on average, with the divergence rate measured by a positive value of the maximum Ljapunov characteristic exponent (LCE) λ associated with the flow /3/.

In recent years LCEs are widely used to characterize attractors. Especially, transitions from periodic to chaotic motions, as well as such between chaotic motions of different character, can be described quantitatively by their spectrum of LCEs. So it was shown by HUBERMAN and RUDNICK /4/ that for the map (1) the emergence of a positive value for the envelope of λ as the control parameter r exceeds the onset value r_∞ takes place in a steep and continuous fashion described quantitatively by a power law

$$\lambda = \lambda_0 (r - r_\infty)^t , \quad (2)$$

where λ_0 is a constant of order unity and t an exponent which is given by

$$t = \log 2/\log \delta = 0.4498...$$

As was first reported by UEDA /5/ there are transitions, however, where the character of attractors changes more rapidly. GREBOGI, OTT and YORKE /6/ have shown that the onset of chaos can occur suddenly even in the map (1), and assigned the term "crisis" to such an event. As the degree of chaos changes considerably at the crisis point, one would expect a sharp increase of the maximum LCE.

3. Universal Properties of Multidimension Dissipative Systems

An interesting question is whether universal behaviour occurs also in two and more dimensional dissipative systems. Concerning the Feigenbaum numbers δ and α there are strong arguments that they are the same in multidimensional dissipative systems, and the values obtained numerically or experimentally are close to those given by the one-dimensional theory (see, e.g. /7-9/).

In order to describe the behaviour of the LCEs for multidimensional dissipative systems, and, especially, to check the relation (2), we have calculated the spectrum of LCEs for different types of motion of a parametrically excited pendulum described by the equation

$$\ddot{x} + \beta x + (1 + A \cos \Omega t) \sin x = 0 \quad (3)$$

and a kicked rotator leading to the Zaslavski map

$$\begin{aligned} y_{n+1} &= b y_n - A \sin x_n , \\ x_{n+1} &= x_n + \omega + y_{n+1} \mod 2\pi . \end{aligned} \quad (4)$$

For the pendulum the spectrum of the LCEs consists of three quanti-

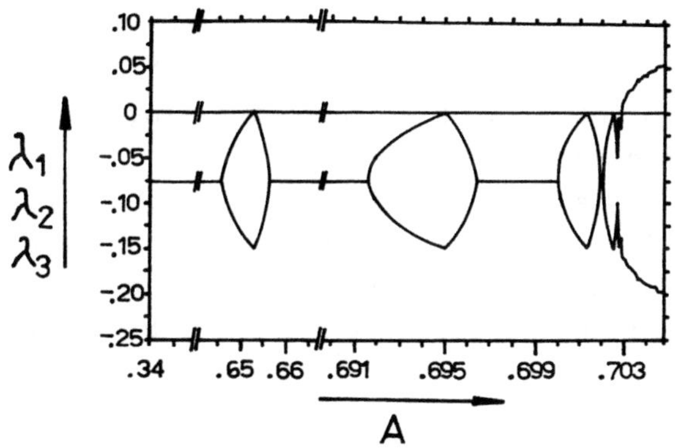

Fig. 1 Oscillating motion with period 2T ($T=2\pi/\Omega$ is the period of the driving force)

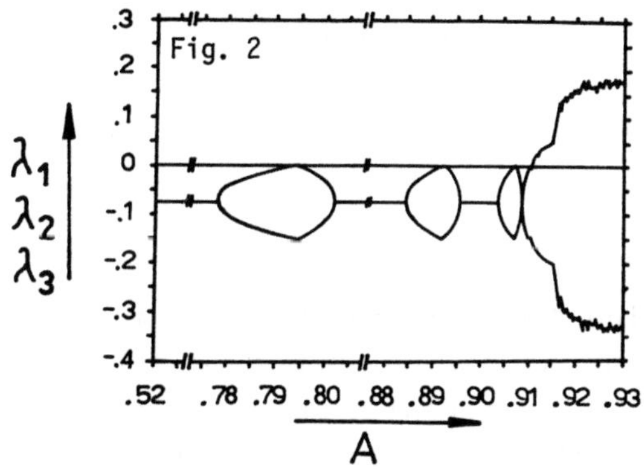

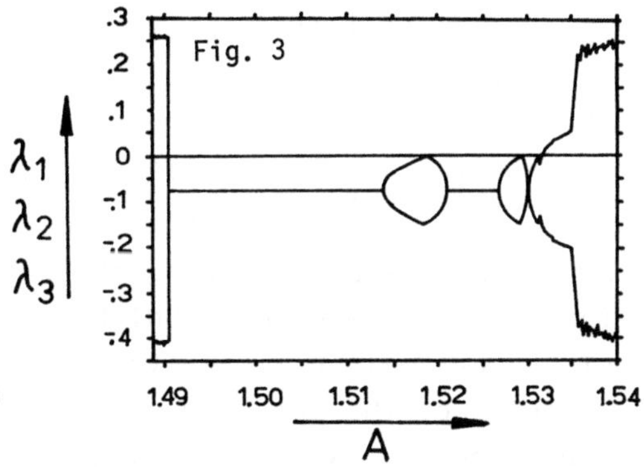

Fig. 2 The evolution of a rotational motion with period T to chaos

Fig. 3 Oscillating motion with period 4T which appears as a "window" in the chaotic regime

ties λ_i (i=1,2,3) with $\sum_i \lambda_i = -\beta/2$ and $\lambda_1 = 0$, $\lambda_2, \lambda_3 < 0$ in the periodic state and $\lambda_1 > 0$, $\lambda_2 = 0$, $\lambda_3 < 0$ in the chaotic state of motion. In the case of the Zaslavski map we have two LCEs with $\lambda_1 + \lambda_2 = \ln b$.

Figures 1-3 show the spectrum of LCEs for the pendulum equation (3) for three parameter intervals where a transition from a periodic regime to chaos takes place via period doubling bifurcations. In all cases the control parameter is the amplitude of the driving force, A, the damping and frequency are fixed at $\beta = 0{,}15$ and $\Omega = 1{,}56$, respectively.

At values A_k where the period doubling bifurcations take place the second LCE approaches zero. The critical value of A where the bifurcation cascade accumulated is $A_\infty = 0.70279...$ For $A > A_\infty$ the motion of the pendulum becomes chaotic with the typical sequence of band mergings known from one-dimensional maps. Hence this is the interval where one would expect a power-law behaviour of the envelope of λ_1.

In both cases the chaotic regime begins with the expected steep and continuous increase of the envelope of the maximum LCE which, however, is followed by a sudden increase of λ_1 characterizing a crisis-like transition to a large chaotic attractor.

Within the interval between A_∞ and the crisis point we find the power-law behaviour of the envelope of λ_1 with values of the exponent t given in the table. The same power law was obtained for an oscillating and a rotational solution of the Zaslavski map (ω =0, b=exp(-2)).

Table 1

Eq.	A_∞	A_ϵ	λ_0	t
(3)	0.70279	(A_∞, 0.705)	1.06 ± 0.2	0.48 ± 0.04
(3)	0.91065	(A_∞, 0.915)	0.7 ± 0.3	0.49 ± 0.06
(3)	1.53175	(A_∞, 1.5348)	1.2 ± 0.2	0.47 ± 0.03
(2)	3.91	(A_∞, 4.6)	1.2 ± 0.2	0.48 ± 0.04
(2)	6.107	(A_∞, 6.19)	1.5 ± 0.6	0.46 ± 0.05

From these results we draw the conclusion that the power law (2) found by HUBERMAN and RUDNICK is true for different types of motion of the parametrically driven pendulum and the two-dimensional Zaslavski map. The obtained values for the exponent t are in a very good agreement with the value t = 0.4498... predicted by HUBERMAN and RUDNICK.

An open question up to now is whether the sharp increase of λ_1 beyond the crisis point also can be described by an analytical expression. Some ideas in that direction can be drawn from a recent paper of DAIDO /10/.

References

1. Bambi Hu: Phys.Reports 91, 233 (1982)
2. M.J. Feigenbaum: J.Stat.Phys. 21, 669 (1979)
3. G. Benettin, L. Galgani, J.M. Strelcyn: Phys.Rev. A14, 2338 (1976) G. Benettin, L. Galgani, A. Giorgilli, J.M. Strelcyn: Meccanica 15, 9, 21 (1980)
4. B.A Huberman, J. Rudnick: Phys.Rev.Lett. 45, 154 (1980)
5. Y. Ueda: J.Stat.Phys. 20, 181 (1979)
6. C. Grebogi, E. Ott, J.A. Yorke: Phys.Rev.Lett. 48, 1507 (1982)
7. R.W. Leven, B.P. Koch: Phys.Lett. A86, 71 (1981)
8. J.S. Testa, J. Perez, C. Jeffries: Phys.Rev.Lett. 48, 714 (1982)
9. P. Cvitanovic: Acta Phys.Polon. A65, 203 (1984)
10. H. Daido: Phys.Lett. A108, 233 (1985)

New Aspects of Instabilities and Chaos in Laser Systems

G. Wallis

Akademie der Wissenschaften der DDR, Zentralinstitut für Optik und Spektroskopie, Rudower Chaussee 6, DDR-1199 Berlin, GDR

1. Introduction

The phenomena to be described occur in nearly all types of laser systems, though with large differences in pump intensity: in single mode and multi-mode lasers, in lasers with homogeneously or inhomogeneously broadened lines, in lasers with or without external drive or feedback, in lasers with and without saturable absorbers. As in all other dynamic systems with similar instability behaviour, the observed effects are results of internal selforganization, possibly by interplay with an external triggering. They appear above a second threshold, the first one being the threshold between incoherent emission and stable laser activity. The main characteristics are gradually enhanced: selfpulsing of the laser output under constant laser conditions, regular at first, then irregular and chaotic with growing intensity, or bistability going to multistability, hysteresis included, frequency pulling and locking, phase locking also. What we can expect in the theoretical analysis of the threshold of laser stability and of the phenomena above it, therefore comprehends the whole arsenal of chaos theory applied to all instable dynamic systems with determined asymptotics: Poincaré mappings and their picturesque clouds of points (sometimes of spiral structure in the laser case), interval mappings or area mappings (sometimes specific circle maps or mappings of the complex plane), strange attractors (fractals), the rippled curves of Ljapunov exponents, and the large amount of bifurcation diagrams (comp. the excellent overview of LEVEN and the other instructive topical lectures of this conference).

Concerning the power spectrum, it has been found out that it is useful to determine the routes to chaos (period doublings or subharmonics generation, appearance of incommensurable frequencies and so on) more than to characterize the state of chaos itself.

A guide through the large field of literature on the standard aspects of instabilities and chaos in lasers, as mentioned above, may be learned from /1-3/ and especially from the topical issue of JOSA B (Jan. 1985) /4/. But only minor overlapping analysis may be found concerning types of basic equations and main orbit structures (as compared e.g. with the case of instabilities and chaos in chemical systems /5/), leading to possible classifications of laser systems and phenomena concerning chaos structures.

2. Basic Equations

Recently some new aspects have been added to this. On the one hand there is the use of homoclinic and heteroclinic orbits as well as homoclinic explosions and homoclinic bifurcations, in their relevance to changes from prechaotic to chaotic states or to changes between chaos types (spiral type to screw type e.g.). On the other hand, we have the new method of attractor determination from time series. After years of intensive and sometimes expensive numerical work to grasp the multiplicity and plurality of chaos phenomena, some revival of analytical methods, further developed in the meantime, took place. Punctually the numerical results gave some hints where analytic methods promisingly could be applied.

Of most practical importance in the future probably will be the new concept of determining existence and dimension of fractal attractors from time series of direct experimental data, i.e. without knowledge of the underlying dynamic equations describing the system. This surely will be of great value for concrete, nonidealized laser systems, where the complete system of equations either is not really known or is too complicated even for long computing times on fast machines.

The first generalization of the basic Maxwell-Bloch system, equivalent to the Lorenz system /6/, includes amplitude and phase of the field strength $E \sim X$ and the polarization $P \sim Y$ by choosing X and Y complex, also $r = r_1 + ir_2$ and $a = 1 - ie$ if dispersive effects are relevant too ($Z+1 \sim$ inversion of the atom system):

$$\left.\begin{aligned} \dot{X} &= -\sigma X + \sigma Y \\ \dot{Y} &= rX - aY - XZ \\ \dot{Z} &= -bZ + r^{1/2}(X^*Y + XY^*) \end{aligned}\right\} \dot{X} = F(X;\mu) \tag{1}$$

having the rotational symmetry

$$R(\alpha)\ F(X,\mu) = F(R(\alpha)X;\mu) \tag{2}$$

with the rotation matrix $R(\alpha)$. $R(\beta) = R(\alpha+\beta)$, replacing the nonzero fixed points of the real equation by a limit cycle (comp. /7,8/).

From this, one has the impression that all the complicated bifurcation structure of the real Lorenz equation, as well as its strange attractor, filling a book /9/ not even exhausting the whole richness of the phenomena in the full 3D space of the parameters σ, r and b, may be pronounced by the truncation of the rotational invariance of the complex Lorenz equation to the mirror symmetry on the real plane $(X,Y) \rightarrow (-X,-Y)$.

Equation (1) gives a system of five real ODEs, which may be reduced again to three real ODEs, but different from the real Lorenz equation, if the adiabatic approximation for the polarization $y_1 = y_2 = 0$ is allowed ($y_1 = (r-z)x_1$, $y_2 = (r-z)x_2$):

$$\begin{aligned} \dot{x}_1 &= -\sigma(1-r+z)x_1 & \quad E &\rightarrow (x_1 + ix_2) \\ \dot{x}_2 &= -\sigma(1-r+z)x_2 & \quad P &\rightarrow (y_1 + iy_2) \\ \dot{z} &= -bz + (r-z)(x_1^2 + x_2^2) & \quad D &\rightarrow (1+z) \end{aligned} \tag{3}$$

or with $\xi = x_1 + x_2 \sim |E|^2 = I$:

$$\begin{aligned} \dot{\xi} &= 2\sigma(r-1) - 2\sigma\xi z \\ \dot{z} &= r\xi - bz - \xi z\ . \end{aligned} \tag{4}$$

This 2D system shows that by adiabatic elimination of the polarization, chaos and irregularities are eliminated too. If they appear in this approximation they are originated by additional effects, as external feedback or signal injection and so on.

3. Attractor Determination from Time Series - Applications to Laser Chaos

As in other dynamic systems, there may arise also in active optical devices the case that only the time series of one single system quantity (property) $x_o(t)$, mostly the laser output intensity I (emitted radiation), is known.

What is proposed consists in the possibility to describe the evolution of the system behaviour in an F-dimensional (FD) phase space, i.e. by a system of ODEs, F in number. Not known is the connection between the measured $x_o(t)$ and the orbit $Y(t) = \{y_1(t)\ldots y_F(t)\}$ in the phase space. What may be hoped, nevertheless, is to give some estimates for the asymptotic behaviour by determining the existence and dimension $d < F$ of an attractor. The procedure developed for this in /10-14/ works as follows: We go from t to $\{t_1 \ldots t_i \ldots t_N\}$, from $x_o(t)$ to $x_o(t_i)$, choose some τ and build the vectors $X_i = \{x_o(t_i), x_o(t_i+\tau)\ldots x_o(t_i+(n-1)\tau)\}$ of them. They span a substitute nD phase space, possibly embedding the attractor. Again the connection between the $X_i^{(n)}$ and $Y(t_i)$ is not known. So we are in a situation which falls into the category of attempts to say something interesting about chaotic systems with very incomplete information.

To give measures of order for chaotic systems, which looks like a contradiction in terms at the first glance, may be realized by invariant probability measures. However, they have the property, to give countable series of such measures instead of only one distinct. This may be done either by a series of generalized entropies K_q (RENYI entropy) of the order q /15-18/ or of generalized dimensions D_q /13,17,19-21/. Both series are of similar structure, analogous to the moment method in kinetics:

$$K_q = \lim_{b,\tau \to 0} \frac{1}{N\tau} \frac{1}{q-1} \log Z_q(b) \qquad D_q = \lim_{b \to 0} \frac{1}{q-1} \log Z_q(b)/\log b$$

$$Z_q(b) = \sum P^q(i_1 \ldots i_N) \qquad Z_q = \sum_{j=0}^{H(b)} P_j^q, \quad P_j^q = \lim_{N \to \infty} N_j/N$$

$P(i_1 \ldots i_N)$ = joint probability $\quad N_j$ = number of attractor points in box j (of length b)

The special expressions for $q \to 0^+$, $q \to 1^+$ and $q = 2$ are put together in Tab. 1. $K_1 = 0$ for regular trajectories, $K_o > 0$, $K_o > K_1$ is a necessary but not sufficient condition for chaos /16/. The "prediction time" T is given by $T \sim K_1$. $K_2 > 0$ is a sufficient chaos condition /16/.

The topological dimension D_T is always integer according to our usual picture of space. For D_q, in general not integer, we have $D_q \geq D_{q+1}$, where equality only holds for "homogeneous" attractors /12,13/. The box-counting algorithms used for these dimensions, converge very slowly for $D_o < 2$ and are not useless for $D_o > 2$ /23,24-26/. But the convergence is very effective for all values of D_2, also for time series of relatively small data points ($\gtrsim 500$). Concerning more detailed studies of this compare /27/. The method of attractor dimension determination mentioned above is guided by the formula in Tab.1 for $D_2^{(n)}$, with $X_i^{(n)}$ replacing $\underline{X}_i$ in the expression for $C_N(b)$, which gives $C_N^{(n)}(b)$. Plotting $\log C^{(n)}(b)$ against log b and resulting in a straight line for a sufficiently long part of the curve, one gets a correlation dimension $D_2^{(n)}$ not depending on b in that region. It may be an open question, whether or how good this constant value represents that to be calculated from the lim $b \to 0$, $N \to \infty$. Changing the embedding dimensions n we come to the question whether there exists a value n* so that $D_2^{(n^*)} = D_2^{(\infty)}$ for $n > n^*$.

Tab. 1 Invariant probability measures of chaotic dynamics

def. $J : \sum_{i=1}^{k} \lambda_i \begin{cases} > 0 & \text{for } k = J \\ < 0 & \text{for } k = J+1 \end{cases}$

λ_i = LYAPUNOV exponents, $\lambda_\alpha^+ > 0$

$\underline{x} = \{x_1 \cdots x_F\}$ F = number of degrees of freedom

= dimension of phase space

<u>$D_0 (= D)$ = fractal dimension</u> (HAUSDORFF, MANDELBROT)

$$= \lim_{b \to 0} \log M(b) / \log b$$

$$= J + \sum_{i=1}^{J} \lambda_i / \lambda_{J+1}$$

(KAPLAN – YORKE conjecture)

M(b) = number of boxes containing attractor points $\sim b^{-D_0}$ (without limes)

<u>$K_1 (= K)$ = KOLMOGOROV entropy</u>

$$= - \lim_{\substack{b, \tau \to 0 \\ N \to \infty}} \frac{1}{N\tau} \sum_{i_1 \cdots i_N} P(i_1 \cdots i_N) \log P(i_1 \cdots i_N)$$

$$= \int d^F x \, \varrho(\underline{x}) \sum_\alpha \lambda_\alpha^+(\underline{x}) \quad \text{(PESIN)} \quad /22/$$

$\varrho(\underline{x})$ = invariant attractor density

<u>$D_1 (= \sigma)$ = information dimension</u>

$$= \lim_{b \to 0} (S(b) / \log b), \quad S(b) = - \sum_{j=0}^{M(b)} p_j \log p_j$$

$$\approx S_0 - D_1 \log b$$

<u>K_2 = correlation entropy</u>

$$= \lim_{\substack{b, \tau \to 0 \\ N \to \infty}} K_{2,N}^{(b)}, \quad K_{2,N}^{(b)} = \frac{1}{\tau} \log \left(C_N^{(b)} / C_{N+1}^{(b)} \right)$$

$$C_N^{(b)} = \sum_{j=0}^{M(b)} p_j^2 \quad = \frac{1}{N^2} \sum_{i \neq j} \Theta \left(b - |x_i - x_j| \right)$$

Θ = step function

<u>$D_2 (= \nu)$ = correlation dimension</u>

$$= \lim_{b \to 0} \log C(b) / \log b, \quad C(b) = \lim_{N \to \infty} C_N^{(b)} = \text{correlation integral}$$

$c(b')$ = standard – two point correlation function

$$= \int_0^b d^F b' c(b') \sim b^{D_2} \text{ (without limes)}$$

$$= \sum_{j=1}^{M(b)} p_j^2$$

This method has been applied to a GaAs diode laser with external feedback /28/, using a time series of 512 data points. Our results show the correlation dimension $D_2^{(n)}$ versus unbedding dimension n, for laser diode output power series of different regularity in time, $D_2^{(\infty)}$ in the range 2.1 to 2.6. From this one can postulate the existence of an attractor describing the asymptotic behaviour of the laser system. In contrast to similar results for hydrodynamic and other systems /14,29-31/ where $n^* < 2D_2^{(\infty)}+1$, in laser systems /27,32-34/, however, we have $n^* > 2D_2^{(\infty)}+1$. This may be caused by different transversal irregularities of the attractor for the complex valued laser fields.

4. Conclusion

Laser systems are seemingly useful to investigate bifurcation schemes as function of the number and structure of ordinary differential equations in the basic system, as well as the unbedding problem in determining the strange attractor of the asymptotic state.

References

1. N.B. Abraham: Laser Focus, May 1983, p. 73
2. R.G. Graham: Chaos in simple laser systems, in: Coupled nonlinear oscillations (Eds.: J. Chandra, A.C. Scott) (North-Holland Publ. Comp., Amsterdam 1983; Springer, Berlin Heidelberg New York 1984)
3. L. Mandel, E. Wolf (Eds.): Coherence and quantum optics (Plenum Press, New York 1984)
4. N.B. Abraham, L.A. Gugiato, L.N. Narduci (Eds.): Instabilities in active optical media, J.Opt.Soc.Amer. 2B, 1 (1985)
5. O. Gurel, D. Gurel: Types of oscillations in chemical reactions, in: Topics in Current Chemistry, Vol. 118 (1984) and Akademie-Verlag, Berlin
6. H. Haken: Synergetics, an Introduction, 2nd ed. (Springer, Berlin Heidelberg New York 1978)
7. J.D. Gibbon: In: Chaos and Order in Nature (Springer, Berlin Heidelberg, New York 1981)
8. A.C. Fowler, J.D. Gibbon, M.J. McGuinnes: Physica 4D, 139 (1982); 5D, 108 (1982); 7D, 126 (1983)
9. C. Sparrow: The Lorenz equations: Bifurcation, Chaos, and Strange Attractors (Springer, Berlin Heidelberg New York 1982)
10. N.H. Packard, J.P. Cruchfield, J.D. Farmer, R.S. Shaw: Phys.Rev. Lett. 45, 712 (1980)
11. F. Takens: in: Lecture Notes in Mathem. Vol. 898 (1981) (Eds.: D.A. Rand, L.S. Young)
12. P. Grassberger, I. Procaccia: Phys.Rev.Lett. 50, 346 (1983); Physica 9D, 189 (1983)
13. H.G.E. Hentschel, I. Procaccia: Physica 8D, 435 (1983)
14. C. Nicolis, G. Nicolis: Nature 311, 529 (1984); Phys.Bl. 41, 5 (1985)
15. A. Renyi: Wahrscheinlichkeitsrechnung (Akademie-Verlag, Berlin 1966)
16. P. Grassberger, I. Procaccia: Phys.Rev. 28A, 2591 (1983)
17. P. Grassberger: Rep.Univers.Wuppertal (1984) WU B 84-31
18. A. Cohen, I. Procaccia: Phys.Rev. 31A, 1872 (1985)
19. H.G. Schuster: Deterministic Chaos (Physik-Verlag, Weinheim 1984)

20. I.D. Farmer, E. Ott, J.A. Yorke: Physica 7D, 153 (1983)
21. P. Grassberger: Physica 10D, 53 (1984); 14D, 365 (1985)
22. Ya.B. Pesin: Usp.Mat.Nauk 32, 55 (1979)
23. J.D. Farmer, E. Ott, J.A. Yorke: Physica 70, 153 (1983)
24. P. Grassberger: Phys.Lett. 107A, 101 (1985)
25. A. Ben-Nizrachi, I. Procaccia, P. Grassberger: Phys.Rev. 29A, 975 (1984)
26. N.S. Greenside, A. Wolf, T. Pignatoro: Phys.Rev. 25A, 3453 (1982)
27. N.B. Abraham et al.: 5th Int.Conf.Lasers and their Applications, Dresden 1985
28. P. Glas, R. Müller, A. Klehr: Opt.Comm. 47, 297 (1983)
29. M. Dubois: Lect.Notes in Physics 164, 177 (1982)
30. B. Malraison, P. Atten, P. Berge, M. Dubois: J.Physique - lettres 44, L-897 (1983)
31. M. Giglio, S. Mosazzi, U. Perini: Phys.Rev.Lett. 42, 2402 (1984)
32. A.M. Albano, J. Abounadi, T.H. Chyba, C.E. Searle, S. Young, R.S. Gioggia, N.B. Abraham: J.Opt.Soc.Am. B(2), 47 (1985)
33. G.P. Puccioni et al.: Phys.Rev.Lett. 55, 339 (1985)
34. P. Glas, R. Müller, G. Wallis: 5th Int.Conf.on Lasers and their Applications, Dresden 1985

Complexity of Patterns Generated by One-Dimensional Cellular Automata

R. Mahnke and A. Budde

Wilhelm-Pieck-Universität Rostock, Sektion Physik, Universitätsplatz 3, DDR-2500 Rostock, GDR

1. Introduction

Cellular automata are mathematical models for complex systems in which many simple components act together to produce complicated patterns of behaviour. These automata serve as key models to study the origins of chaos (randomness) in physical systems. There are examples in which a cellular automaton evolving from a single initial state produces a structure so complicated that some features of it seem random. WOLFRAM, PACKARD and co-workers have been investigated one-dimensional and two-dimensional automata extensively in several ways /1/. In his "Twenty Problems in the Theory of Cellular Automata" /2/ WOLFRAM formulates in problem 15 a question concerning the randomness of sequences generated by cellular automata.

To calculate the degree of randomness of sequences we use the concept of complexity measures of words /3/, especially the subword complexity. This complexity is a mathematically precise measure to examine the chaotic or regular structure of sequences, the algorithm given in the following is very easy.

There are, of course, several statistical quantities to characterize cellular automata behaviour. Beyond entropies, Ljapunov exponents, correlation functions and others, we believe that the number of subsequences ("blocks") which occur in the pattern is a good candidate to measure quantitatively the degree of randomness and to distinguish different patterns.

In this paper we want to investigate a complete set of patterns generated by one-dimensional cellular automata and discuss the calculated complexities to classify the configurations.

2. Generation of Evolving Patterns

In the notation of one-dimensional cellular automata $a_i^{(t)}$ is taken to denote the value of site (cell) i at time step t. The site values may evolve by iteration of the mapping

$$a_i^{(t+1)} = F(a_{i-r}^{(t)},\dots,a_i^{(t)},\dots,a_{i+r}^{(t)}) \ . \qquad (1)$$

Let us take into account nearest-neighbour interactions (range r=1) only and that the sites are specified by the values 0 (false) and 1 (true), then (1) reads for the elementary case

$$a_i^{(t+1)} = F(a_{i-1}^{(t)},a_i^{(t)},a_{i+1}^{(t)}), \quad \text{with} \quad a_i \in \{0,1\} \ . \qquad (2)$$

Since we have $2^3 = 8$ possible values of the three variables a_{i-1}, a_i, a_{i+1}, we are able to construct in general $2^8 = 256$ distinct cellular automaton rules according to (2). But we restrict ourselves to rules for growing patterns, that means we fix

$$F(0,0,0) = 0; \; F(0,0,1) = 1; \; F(1,0,0) = 1. \tag{3}$$

These restrictions (3) (quiescence and boundary conditions) leave us $256/2^3=2^5=32$ possible local rules for the generation of growing triangle-like patterns. In /4/ other forbiddances of rules were studied, particularly the so-called reflection symmetry condition which guarantees isotropy in cellular automaton evolution.

In Fig. 1 the evolution of a whole set of one-dimensional elementary cellular automata according to 32 different rules is shown. The generation always starts from an initial configuration containing a single site with value 1 in analogy to the growth of a crystal from a microscopic seed. The time evolution is shown for 60 time steps (value 1 represented by stars, value 0 by blanks) where in each step a sequence (0-1 word) determined by the specific local rules (2) is generated. From Fig. 1 a considerable variety of behaviour, and the fact that different rules may generate identical patterns, are evident. In order to identify each configuration we use an eight-digit binary number for (2) together with (3) and take its decimal representation as identification number in Table 1. To make it more transparent, we give one example applied to the top-left pattern of Fig.1:

$$\left.\begin{array}{l} F(0,0,0) = 0 \\ F(0,0,1) = 1 \\ F(0,1,0) = 1 \\ F(0,1,1) = 1 \\ F(1,0,0) = 1 \\ F(1,0,1) = 0 \\ F(1,1,0) = 1 \\ F(1,1,1) = 1 \end{array}\right\} \begin{array}{l} \text{rule number} \\ 0\cdot 2^0 + 1\cdot 2^1 + 1\cdot 2^2 + 1\cdot 2^3 \\ + 1\cdot 2^4 + 0\cdot 2^5 + 1\cdot 2^6 + 1\cdot 2^7 \\ = 222 \text{ (decimal code)} \end{array} \tag{4}$$

3. Calculation of Subword Complexities

Following the general ideas of KOLMOGOROV and CHAITIN we introduce the term "complexity" of an object in the sense that the memory or time needed to generate the object gives us a measure (a real number) to characterize the object as simple or complex. This so-called algorithmic complexity is based on the fact that simple objects have a great compressibility, that means the length of the algorithm to generate them is short, but on the other hand for complex systems, where the degree of compressibility is small (or zero) we need a long program to generate them. In earlier papers it was proposed to apply the concept of complexity to the analysis of biopolymers /3,5/.

Among the different possibilities to measure the complexity of a pattern (sequence, word, string) here we choose the subword complexity. This mathematically well-defined function, proposed by THIELE and HEINZ /6,7/ and applied to biopolymers by EBELING and MAHNKE /8-10/, may serve as the basic quantity for the classification of the generated pattern (Fig. 1).

The following notations familiar from formal language theory will be used. Let $X = \{X_1, X_2, \ldots, X_n\}$ a finite alphabet with $|X| \equiv \mathrm{card}X = n$ elements or building stones (in our case n=2, for DNA n=4, for proteins n=20) $p, q \in X^*$ words (sequences) constructed over X and $l(q)$ length of sequence q. The subword property "p is subword of q" (p in q) is defined as follows

$$p \text{ in } q: = \exists q_1 \exists q_2 \; (q_1, q_2 \in X^* \wedge q = q_1 p q_2) \,. \tag{5}$$

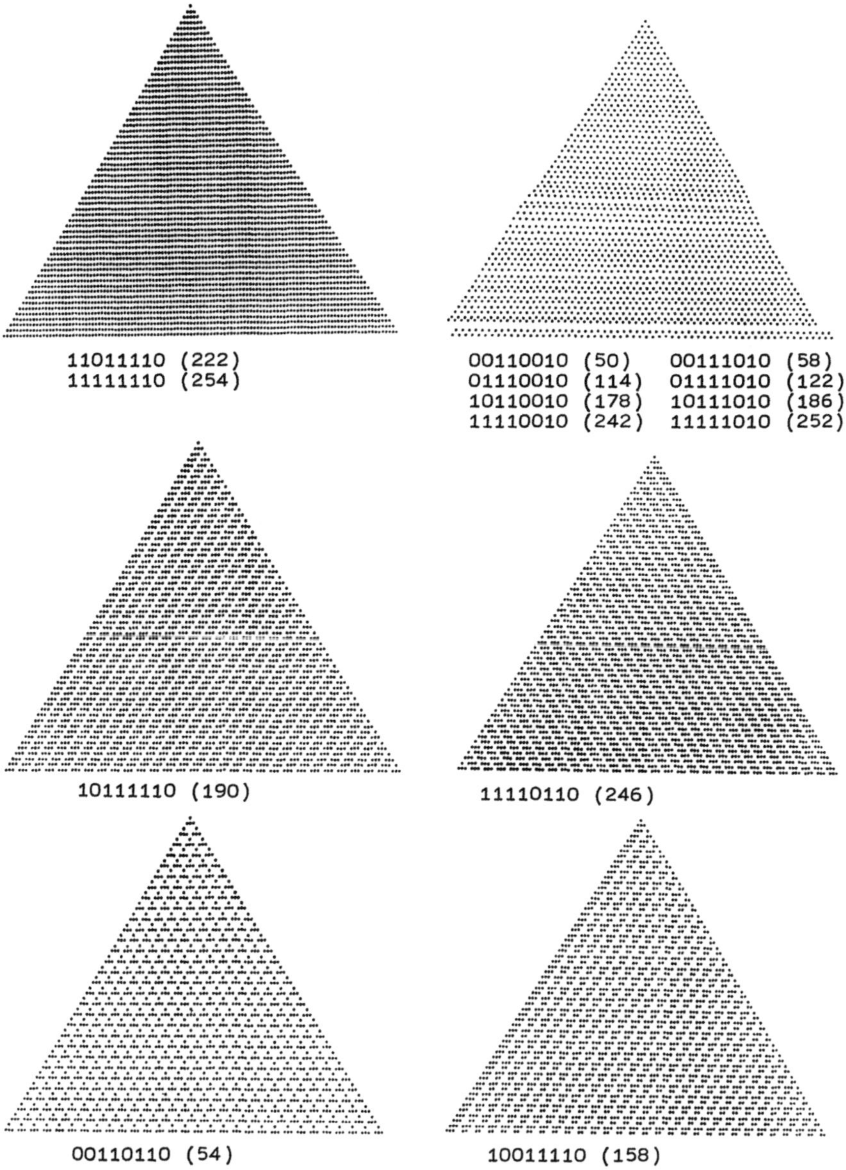

Fig 1 Evolution of one-dimensional cellular automata according to the 32 possible rules. Detailed explanation see text.

Let A(i,q) the set of all subwords of q with length i

$$A(i,q) = \{p | p \text{ in } q \wedge l(p) = i\} \qquad (6)$$

and the size of A(i,q) (number of all subwords of length i) denoted by

$$f(i,q) = |A(i,q)| \quad \text{for } i=1,2,\ldots,l(q) . \qquad (7)$$

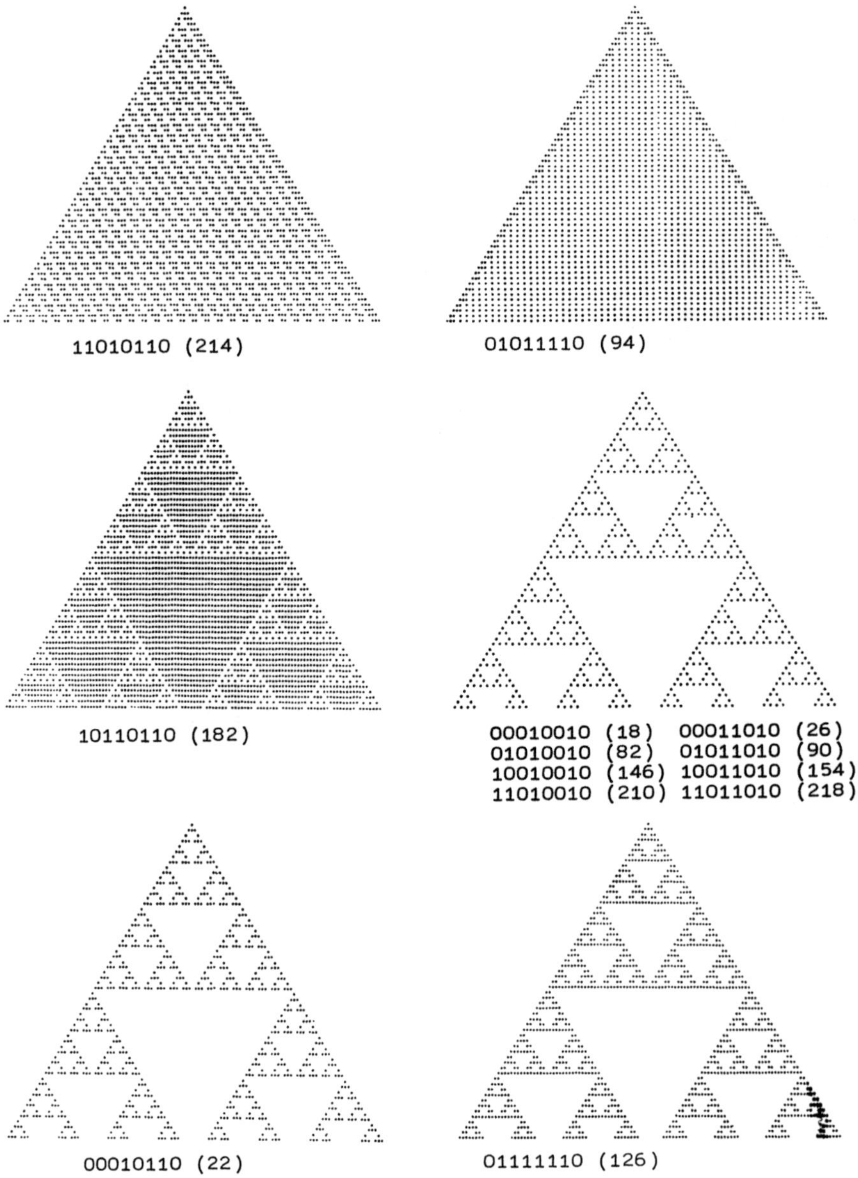

Fig. 1 Continued

Then we define four measures of the complexity of a sequence q which are connected with the overall number of subwords occurring in q. The simplest measure is the subword complexity /6/

$$K(q) = \sum_{i=1}^{l(q)} f(i,q) \quad . \tag{8}$$

From (8) we can derive the divided subword complexity /7/

$$D(q) = K(q)/l(q) \quad . \tag{9}$$

10010110 (150)

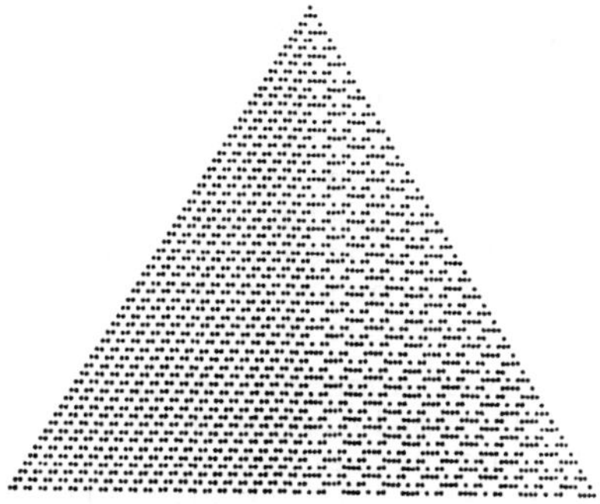

00111110 (62)

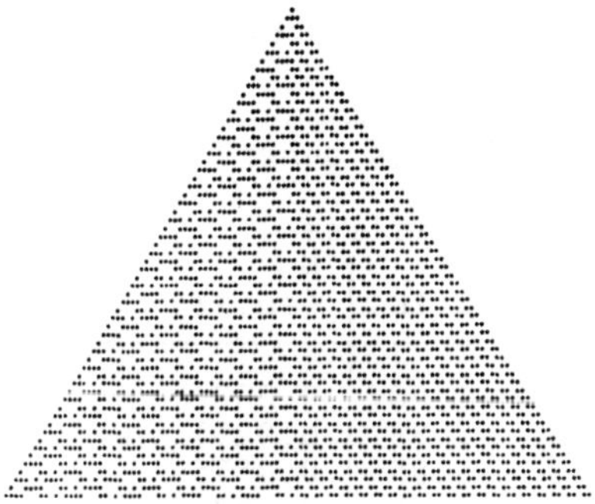

01110110 (118)

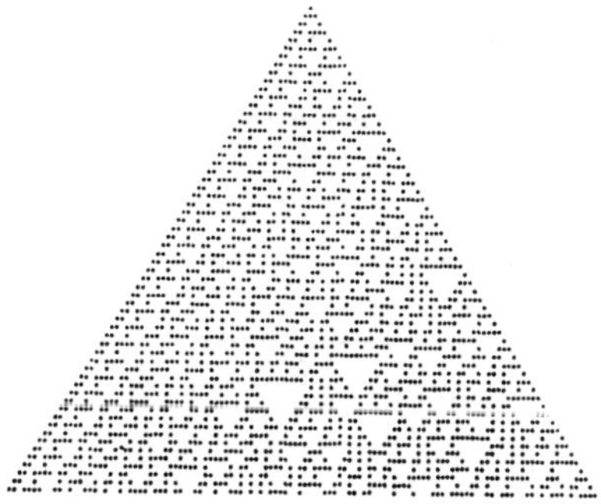

00011110 (30)

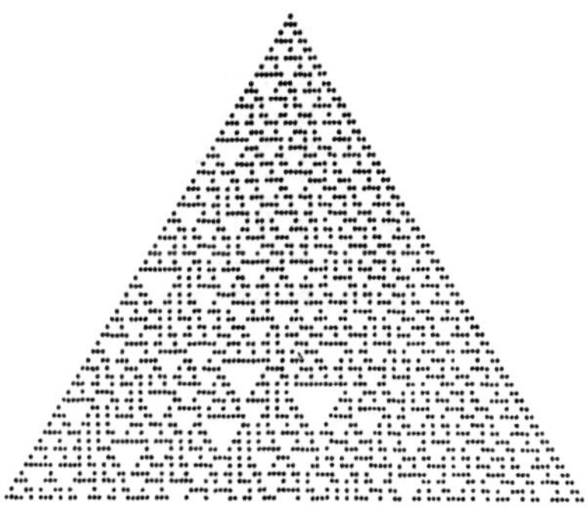

01010110 (86)

The set of 32 rules produces 17 different patterns, of which 4 are mirror-symmetric.

Fig. 1 Continued

Furtheron let us introduce the modified subword complexity /8/

$$M(q) = (2K(q) - l(q))/l(q) \tag{10}$$

and the normalized subword complexity /11/

$$N(q) = (2K(q) - l(q))/l(q)^2 . \tag{11}$$

From the mathematical point of view we only mention the upper and lower bound of K(q)

$$l(q) \leqslant K(q) \leqslant l(q) \cdot (l(q)+1)/2 \qquad (12)$$

and refer to a recent paper by HUNYADVARI and IVANYI /11/ concerning the ~~proteins~~ of the functions (8)-(11). Using a computer program (given in detail elsewhere /9,10/) for calculating the subword complexities of the generated pattern, the results are given in Table 1.

From the analysis presented in Table 1 we conclude that the distribution of patterns over the complexity (e.g. over the normalized value N(q) (11)) passes from very simple spatially homogeneous states ($N \sim 10^{-2}$) over some periodic or symmetric structures ($N \sim 0.2 \ldots 0.6$) to very complicated aperiodic behaviour ($N \sim 0.9$). In agreement with /1/ the sequences can be grouped into general classes and the proposed measures of complexity answer the question concerning the randomness of sequences.

Table 1 Subword complexities of the patterns generated by cellular automata (comp. Fig. 1)

Decimal identi- fication code	*	Total number of sub- words	Analysis of the sequences generated in the 60th time step Subword complexities			
			K (eq.8)	D (eq.9)	M (eq.10)	N (eq.11)
222	2	3600	119	1.00	1.00	0.0084
50	8	7140	237	1.99	2.98	0.0251
190	2	13866	467	3.92	6.85	0.0575
54	1	13866	467	3.92	6.85	0.0575
158	2	15548	581	4.88	8.76	0.0736
94	1	17114	691	5.81	10.61	0.0892
182	1	59420	903	7.59	14.18	0.1191
18	8	62122	1721	14.46	27.92	0.2346
22	1	62554	1731	14.55	28.09	0.2361
126	1	79638	4031	33.87	66.75	0.5609
150	1	81948	4207	35.35	69.70	0.5858
62	2	88916	4129	34.70	68.39	0.5747
30	2	128679	6476	54.42	107.84	0.9062

* Number of different rules generating identical or mirror-symmetric patterns

References

1. D. Farmer, T. Toffoli. S. Wolfram (Eds.): Cellular Automata; Proceeding of an Interdisciplinary Workshop, Physica 10D, (1984) no. 1/2
   S. Wolfram: Nature 311, 419 (1984)
   N.H. Packard, S. Wolfram: J.Stat.Phys. 38, 901 (1985)
2. S. Wolfram: Physica Scripta T9, 170 (1985)
3. W. Ebeling, R. Feistel: Physik der Selbstorganisation und Evolution (Akademie-Verlag, Berlin 1982)
4. S. Wolfram: Rev.Mod.Phys. 55, 601 (1983)
5. W. Ebeling, M.A. Jimenez-Montana: Math.Biosci. 52, 53 (1980)
6. H. Thiele: Zur Definition von Kompliziertheitsmaßen für endliche Objekte, in: Organismische Informationsverarbeitung (Ed.: F. Klix) (Akademie-Verlag, Berlin 1974)
7. M. Heinz: Elektr.Inform.u.Kybern. (EIK) 13, 27 (1977)
8. W. Ebeling, R. Mahnke: studia biophysica 17, 173 (1978)
9. R. Mahnke: Wiss.Zeitschr.WPU Rostock, MNR 27, 931 (1978)

10. R. Mahnke: Zur Komplexität und Evolution biologischer Makromoleküle, Thesis, Wilhelm-Pieck-Universität Rostock 1980
11. L. Hunyadvari, A. Ivanyi: On some complexity measures of words, in: Proc.of Conf. Automata, Languages and Mathematical Systems, Budapest, Karl-Marx-Universität, 1985, p. 67-82

Index of Contributors

MIX
Papier aus verantwortungsvollen Quellen
Paper from responsible sources
FSC® C105338

If you have any concerns about our products,
you can contact us on
ProductSafety@springernature.com

In case Publisher is established outside the EU,
the EU authorized representative is:
Springer Nature Customer Service Center GmbH
Europaplatz 3, 69115 Heidelberg, Germany

Printed by Libri Plureos GmbH
in Hamburg, Germany